'There have been other Tesla biographies, but this is the one I have been waiting for. Neither hagiographic nor hatchet-job, it sets its mercurial subject in his cultural and historical context: a visionary and showman, part genius and part crank, totally a product of his age. Tesla cannot be understood without a clear view of the (uniquely American) legend he embedded himself within, and Iwan Rhys Morus expounds that view brilliantly. Tesla, he shows us, was – like his one-time boss and rival Thomas Edison – inventing nothing less than the electrified future.'

—Philip Ball, author of *Invisible: The Dangerous Allure of the Unseen*

'Nikola Tesla saw himself as a rebel, a free-thinker, a disruptor and the sworn enemy of scientific mediocrity. Brilliant in his experimentation, chaotic in his methodology, Tesla was the twentieth century's first visionary tech entrepreneur.'

—Thomas Dolby, musician and sound tech pioneer

'Crisply succinct, beautifully synthesized.'

—*Nature*

ABOUT THE AUTHOR

Iwan Rhys Morus is professor of history at Aberystwyth University. He graduated in Natural Sciences from Cambridge and completed his doctorate there in the history and philosophy of science. He has published widely on the history of science. Recent publications include *Michael Faraday and the Electrical Century* (Icon Books, 2017) and *The Oxford Illustrated History of Science* (Oxford University Press, 2017).

NIKOLA TESLA
AND THE
ELECTRICAL
FUTURE

IWAN RHYS MORUS

ICON

First published in the UK in 2019
by Icon Books Ltd, Omnibus Business Centre,
39–41 North Road, London N7 9DP
email: info@iconbooks.com
www.iconbooks.com

This edition published in the UK in 2020 by Icon Books Ltd

Sold in the UK, Europe and Asia
by Faber & Faber Ltd, Bloomsbury House,
74–77 Great Russell Street,
London WC1B 3DA or their agents

Distributed in the UK, Europe and Asia
by Grantham Book Services,
Trent Road, Grantham NG31 7XQ

Distributed in the USA
by Publishers Group West,
1700 Fourth Street, Berkeley, CA 94710

Distributed in Australia and New Zealand
by Allen & Unwin Pty Ltd,
PO Box 8500, 83 Alexander Street,
Crows Nest, NSW 2065

Distributed in South Africa
by Jonathan Ball, Office B4, The District,
41 Sir Lowry Road, Woodstock 7925

Distributed in India by Penguin Books India,
7th Floor, Infinity Tower – C, DLF Cyber City,
Gurgaon 122002, Haryana

Distributed in Canada by Publishers Group Canada,
76 Stafford Street, Unit 300
Toronto, Ontario M6J 2S1

ISBN: 978-178578-617-4

Typeset in Adobe Caslon by Marie Doherty

Printed and bound in Great Britain
by Clays Ltd, Elcograf S.p.A.

Contents

Part 5: Visions of Tomorrow

Prologue

Nikola Tesla is in many ways one of the most enigmatic, curious, and controversial figures in the history of science. He was controversial during his own lifetime too, with opinion divided between those who thought he was either a visionary, a charlatan, or a fool. Whatever he was, he was full of apparent contradiction. Tesla was a consummate showman and a very private recluse – a man of science who seemed to be addicted to self-promotion and sensationalism. He was a prolific inventor of technologies that sometimes helped make other men's fortunes, but failed completely to make one for himself. Tesla embodied the aspirations and the paradoxes of an age of innovation that seemed to have the future firmly in its grasp. Looking back now at the future that Tesla imagined, it still seems very familiar. It is no coincidence that Elon Musk named the company making his electrical car of the future 'Tesla'. Tesla's *fin-de-siècle* invocation of an electrical future still casts its shadow over the ways we understand our future now.

The final decades of the nineteenth century and the opening decade of the twentieth century were a time of unprecedented technological transformation. Tesla was one of the key innovators of that innovative age. He was a key figure in developing new ways of generating and transmitting electrical power. He played a vital role in establishing the networks of power that still run our economies more than a century later. The last quarter of the nineteenth century saw the global spread of the telegraph network, the invention of the telephone, and the beginnings of wireless telegraphy. The discovery of X-rays and radioactivity seemed to open up new vistas for the future. The beginning of the twentieth century saw the beginnings of powered flight. This was the world that produced Tesla, and that he helped produce. His restless speculations and experiments about the way tomorrow would be also helped inaugurate new ways of trying to understand the future – the ways we still turn to as we try to make sense of the future now, and imagine how to get there.

Tesla is often described (and some of his promoters described him in this way during his own lifetime) as a man ahead of his time. It is a relatively common way of talking about great innovators – Leonardo da Vinci is another example of someone often talked about in this way. In Tesla's case, however, the suggestion could not be more wrong. He was in all ways the product of the scientific and technological turmoil of those final decades of the nineteenth century during which the future seemed to be coming closer on an almost daily basis. That is why this book will not be just a biography of Nikola Tesla. It will take the inventor as its guide and follow him through the cut-throat entrepreneurial culture of late Victorian and Edwardian electrical invention. I want to explore the electrical future that Tesla saw himself creating, and the raw materials from which that future was being forged.

The decades during which Tesla completed his education, came to America and made himself as an inventor were characterized by a near-addiction to innovation. Both the Old World that Tesla left, and the New World where he lived for most of his adult life were undergoing rapid technological change. Electricity was no longer just a thing of lecture theatres and exhibitions, but was leaking everywhere into everyday life. Cityscapes and townscapes across Europe and America were now festooned with the paraphernalia of electrical technology. Cables criss-crossed the sky, carrying power to homes and businesses, or buzzing with information from telegraphs and telephones. Electric tramcars trundled down the streets. At night, public buildings and shopfronts dazzled the eye with spectacular displays of electric lighting. To many people, it looked as if tomorrow's world had already arrived, and people like Tesla were the ones who had delivered it.

This was a time when engineers and inventors could be heroes. They were held up and celebrated as examples of what discipline and perseverance could achieve. As we shall see, even on the far frontiers of the Austro-Hungarian Empire it was possible for a young man like Tesla to dream of a life of invention. Invention offered a road to fame and fortune. More importantly, it seemed to offer an opportunity to change the world. The prevailing image of invention and inventors was one of determined individualism. Tesla, of course, fashioned himself carefully to fit that image, as did others like Thomas Edison. To be an inventor Tesla had to be a showman too. Electricity seemed made for spectacle and Tesla turned himself into a consummate performer of electrical spectacle. He quite literally made himself part of the display, making it seem as if he held the electrical future in his hands.

Electricity and invention at the beginning of the twentieth century were bound up with imagining the future. The nineteenth century was in many ways the century that saw the invention of the future. The nineteenth-century future was imagined as a place that would be different from the present and generated by technological progress and innovation. This was particularly the case with electricity. Electricity was understood as the stuff from which the future would be made, to such a degree that it was almost impossible to discuss electricity at all without invoking its future. To most people, electricity simply was instantaneous communication, with or without wires; new sources of power; locomotion; improved agriculture and control of the weather; weapons and flying machines – and electrically generated health. This was a future extrapolated from bits and pieces of present technology. One of the things that made Tesla so successful was that he was very good at offering compelling images of the future his inventions would deliver.

The stories that Tesla told about the future of wireless communication and wireless power that he was trying to build at Wardenclyffe during the first few years of the twentieth century were very much part of a wider culture of speculating about the future – both in fact and in fiction. The magazines and newspapers that reported on Tesla's activities and relayed the visions of the future that he promised were filled with such stuff. They published scientific romances by the likes of H.G. Wells as well – and sometimes the scientific romance and the scientific facts were hard to disentangle. It was no coincidence that Tesla's speculations about wireless communication with Mars, for example, were aired at the same time as everyone was reading *The War of the Worlds*. Telling tales about the future was integral to the business of invention. Inventors like Tesla needed to excite the public's imagination to sell them their vision of the future – and to attract the attention of the money men whose cash would be needed to make those visions real.

One of the reasons, I think, why Tesla remains such a fascinating and seductive figure is that the image of invention he wrapped around himself remains a very familiar one. That image – which Tesla helped create – of the inventor as an exceptional kind of individual is still the way we tend to think about inventors and innovation today. Just as people in the nineteenth century associated invention with individuals, so do we. Where they made Brunel, or Edison, or Tesla into heroes, we do the same with their contemporary equivalents – though maybe we are a little more inclined than our great-grandparents to make them villains instead. Whether we think of them as heroes or as villains though, the image of invention that we have inherited from the nineteenth century is one that portrays innovation as something that is produced through the exertions and talents of remarkable individuals. Tesla's self-portrait as the uniquely gifted, asocial, daydreaming obsessive is one that still resonates for us today.

That notion of invention as belonging to exceptional individuals like Tesla has its impact too on the way we tend to think about the future and how – and by whom – it gets made. As far as Tesla was concerned, the future was going to be made by him. Tesla's future seems familiar to us, not because we are living in it, but because we still imagine the future in the same sort of way and using the same sorts of ingredients as he did. We make our futures out of bits and pieces of the present. That is one reason, at least, why we should take Tesla seriously still. In the rest of this book I use Tesla as a way in to the future as it looked from the end of the nineteenth century. Understanding that future is important in my view, because it helps us understand why we now imagine the future as we do. Understanding Tesla can tell us not just about how we got to where we are, it can help show us why we think as we do about where we are going, too.

THE ELECTRICAL CENTURY

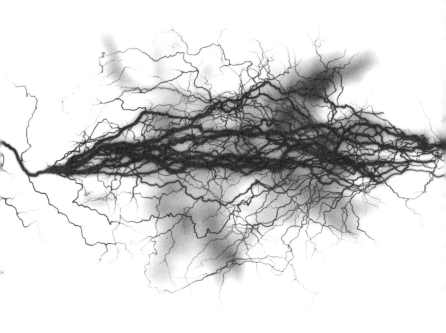

CHAPTER 1

A Child of the Storm

What did the future look like in 1856? For Milutin and Djuka Tesla, living in the village of Smiljan in the province of Lika in Croatia it must have appeared quite uncertain in many ways. They were Serbs, living in Croatian lands, and members of the Orthodox Church in a predominantly Catholic Austrian Empire. Until a few years previously, Milutin had been an Orthodox priest in the town of Senj on the Adriatic coast. The family had moved to Smiljan in 1852, hoping to make a better living in a more prosperous place and with a larger population of co-religionists to support their priest. The times must have seemed precarious, nevertheless. It was only a few years since the Empire had been convulsed by the wave of revolutions that had swept across Europe in 1848. The little town of Smiljan had certainly not been immune to those convulsions, and the unstable frontier with the Ottoman Empire was not far away.[1]

According to family legend, when Nikola Tesla was born at midnight on 9 July by the old Julian calendar, a thunderstorm was raging over the village. The midwife reputedly worried that the infant Tesla would be 'a child of the storm'. Djuka apparently responded, 'No, of light.'[2] It is a powerful image, and one that Tesla himself made much of in later life. The story certainly fits in well with the image Tesla wanted to convey of his own unique genius. His special link with electricity had been forged at the moment of his birth.

Whether or not a storm really raged over Smiljan that night, it seems that Tesla was a sickly child. He was baptised immediately, which suggests some concern that he might not survive. Like all male children born in the military frontier, the young Tesla was promptly enlisted in the local regiment, with the expectation that he would commence service at the age of fifteen.

Though Milutin had broken with family tradition by joining the priesthood, the Teslas were a military clan. Nikola's grandfather had served first in Napoleon's army in the Illyrian provinces that had been ceded by Austria to France, then, following Napoleon's defeat, in the Austrian imperial army. Milutin and his brother Josif had both been enrolled as students in the Austrian Military Officers' School until Milutin had rebelled and decided to join the church instead. Tesla's uncle Josif remained in the army and became a mathematics professor at the military academy.

Djuka Mandić on the other hand came from a line of priests. Her father and her grandfather had both entered the church, and her brother Nikolai became the Archbishop of Sarajevo. There were military men on the Mandić side of the family too, with Nikola's uncle Pajo becoming a colonel in the Austrian army. Both families were Serbs, and committed both to the Orthodox Church and to a vision of a future independent Serbian state.

The Tesla household – and the wider family – within which the young Nikola grew up was clearly one that valued the life of the mind. His uncle Josif was not just a professor of mathematics but the author of a number of mathematical works. His father Milutin was prosperous enough to be able to collect a growing library of books, including works in science and mathematics as well as the theological tomes that might be expected in a priest's study. He also wrote regularly for a number of Serbian journals and magazines on a variety of topics, particularly on the need for education in the Serbian language. There were clearly plenty of opportunities for an eager and inquisitive young child. Tesla remembered his mother

Djuka as 'an inventor of the first order' who would 'have achieved great things had she not been so remote from modern life and its multifold opportunities'.[3]

Tesla was exaggerating the remoteness. Smiljan might have been in a relatively obscure corner of the Austrian Empire, but it was still part of a modern European state, and one that during his childhood was changing rapidly in the wake of the 1848 revolution. The political and ethnic geography of this eastern frontier of the empire was certainly complex. Tesla's ancestors would have settled in the area having fled Ottoman encroachments a few centuries earlier. Their settlement on the precarious border was a deliberate imperial policy designed to provide a ready supply of local soldiery to defend against possible Ottoman invasion. Croats, Serbs and other ethnic groups, Catholic, Orthodox, and even Muslim, lived cheek by jowl along the border. When Tesla was ten years old, the old Austrian Empire became the new Austro-Hungarian Empire with the incorporation of the Kingdom of Hungary on 30 March 1867. The world in which he was growing up was rapidly changing, and those changes would create new opportunities and new possible futures for ambitious young men.

Milutin Tesla's own activities as a writer show how fully engaged the family were with the world around them. Milutin contributed regularly to the Novi Sad *Diary* and other Serbian publications. He mainly wrote about the need for Serbian language education, complaining that 'except for the clergy and merchants or tradesmen, here and there, hardly anyone knows how to sign his name in Serbian.' He complained that 'Serbs in Croatia do not have High Schools, Teachers' Colleges, or any other public places of learning.'[4] As well as his journalism, he was involved in campaigns to establish schools for the local population. Smiljan might have been a long way from the empire's centre of political, economic, and technological power in

Vienna, but it was not remote from it. The empire's political leaders were busily embracing a technological future and the tentacles of progress were spreading out across its territories.

Throughout the 1840s the empire's railway networks were spreading rapidly. This was the result of a determined effort by the state to industrialize and innovate. By the 1850s there were railway lines running south from Vienna to Ljubljana in Slovenia and onwards towards Trieste. By the 1860s the railways were encroaching on Croatian territory too, although it would be a long time before they arrived at Smiljan. The railways were a deliberate effort to consolidate the empire. As contemporary commentators noted, railways made the world smaller and more manageable. As the English commentator Dionysius Lardner put it, with the advent of the railways 'the whole population of the country would, speaking metaphorically, at once advance *en masse*, and place their chairs nearer to the fireside of their metropolis by two-thirds of the time which now separates them from it; they would also sit nearer to one another by two-thirds of the time which now respectively alienates them.'[5] In the Austrian Empire, they were an equally deliberate effort to improve industry.

The Austrian Empire's adoption of the electromagnetic telegraph during the 1840s was as calculated and strategic as the enthusiasm for the railways. Telegraphy in the empire was an imperial monopoly – the *Telegraphenregal* – and its adoption was a deliberate attempt to acquire the trappings of a progressive and modern state. Like the railways, the electromagnetic telegraph made the world seem smaller. It annihilated time and space, said its promoters. It certainly revolutionized the speed at which information could travel. For the empire's political leaders, it was another tool in their efforts to subjugate the various ethnic groups that lived within its borders and to create a homogenous and culturally united state. In 1865 the Austrian Empire was one of the signatories of the International Telegraph Convention in Paris. A few years later the International Telegraph Conference assembled in Vienna. The Austrian Empire was ready to embrace the future, and that electrical future was coming closer to Smiljan and to Nikola Tesla.

Tesla himself recalled his own first encounter with electricity from a very early age. As a very young boy he remembered playing with the family cat, Macak. Stroking the animal on a particularly cold and dry evening, he 'saw a miracle that made me speechless with amazement'. The cat's back 'was a sheet of light and my hand produced a shower of sparks loud enough to be heard all over the house'. His father told him that 'this is nothing but electricity, the same thing you see through the trees in a storm.' Tesla found himself wondering if nature was like the cat, and whether God was the one who stroked it to generate lightning. Looking back at the event decades later, Tesla supposed that it must have been the first time that he started thinking about the nature of electricity.[6]

He certainly saw himself as having inherited his mother's gift for invention. He traced his inventiveness to a very early age, describing how he had made his own fish hook to catch frogs. In another experiment, he 'acted under the first instinctive impulse which later dominated me – to harness the energies of nature to the service of man'. He made a machine powered by 'May-bugs' – which seems to have worked well until one of his friends ate the insects. He remembered disassembling – and failing to reassemble – his grandfather's clocks. He also 'went into the manufacture of a kind of pop-gun which comprised a hollow tube, a piston, and two plugs of hemp'.[7] Along with inventiveness came introspection. Tesla said of his own inventive gifts that he had spent so much of his young childhood inside his own head that he had become very good at imagining in detail how things might work.

In 1863, when Nikola was seven years old, the Teslas moved to the slightly larger town of Gospić, following his older brother's tragic death in an accident. His brother Dane died falling from a horse and his death traumatized the young Nikola and his parents. In later life Tesla recalled his brother as being 'gifted to an extraordinary degree – one of those rare phenomena of mentality which biological investigation has failed to explain'. From then onwards, the 'recollection of his attainments made every effort of mine seem dull in comparison'.[8] More immediately, Dane's death meant the end of an idyllic (in memory at least) rural childhood, and the transfer of Milutin's hopes and ambitions for his older son to his younger one.

Gospić, among other things, meant school, and a move to strange and forbidding new surroundings. In principle, at least, schooling within the Austrian Empire was free and compulsory between the ages of six and twelve, so Tesla had already attended school in Smiljan for a year. The school in Gospić offered more opportunities, however. Tesla remembered seeing mechanical models there for the first time, and being inspired to build his own simple water turbines as a result, which he tried out in the local stream. At around the same time he 'was fascinated by a description of Niagara Falls I had perused, and pictured in my imagination a big wheel run by the Falls.'[9] He even told a sceptical uncle (his mathematical uncle Josif perhaps) that he would one day go to America to carry out his scheme.

When he was ten, Tesla moved on to the local gymnasium to continue his schooling. Here he encountered 'various models of classical scientific apparatus, electrical and mechanical'. He remembered that the 'demonstrations and experiments performed from time to time by the instructors fascinated me and were undoubtedly a powerful incentive to invention'.[10] He was coming to excel in mathematics as well. He put this down to the facility for seeing things accurately inside his own head that he thought was at the root of his capacity for invention. He had an 'acquired facility of visualizing the figures and performing the operations, not in the usual intuitive manner, but as in actual life … it was absolutely the same to me whether I wrote the symbols on the board or conjured them before my mental vision.'[11] He was developing an ability to see things and concepts inside his head and examine them from different angles.

He was still obsessed with inventing things – or that is how he remembered his past self looking back from the vantage point of a successful life of invention in the future. He developed an idea for an engine that worked by creating a vacuum, for example. He even built a prototype and was delighted when it seemed to work. Tesla was 'delirious with joy', he remembered. He wanted to use the contraption to power a flying machine. In his imagination he would transport himself through the air to distant places, and now he knew just how to do it in reality: 'a flying machine with nothing more than a rotating shaft, flapping wings, and – a vacuum of unlimited power!' Tesla was devastated when it dawned upon him that his fabulous vacuum engine could not possibly work in the way he had envisaged, and that the slight movement he had observed in his prototype must have been caused – ironically – by a leak.[12]

The next stage of Nikola's education meant leaving home. At the age of twelve he was sent to the gymnasium in Karlovac, about 150 kilometres away from Gospić, living with his father's sister and her husband (another military man) while studying there. Tesla was by now a devotee of electricity, fascinated by the experiments carried out by the school's professor of physics. He remembered in particular 'a device in the shape of a freely rotatable bulb, with tinfoil coatings, which was made to spin rapidly when connected to a static machine'. Tesla wrote later that it was 'impossible for me to convey an adequate idea of the intensity of feeling I experienced in witnessing his exhibitions of these mysterious phenomena. Every impression produced a thousand echoes in my mind. I wanted to know more of this wonderful force; I longed for experiment and investigation and resigned myself to the inevitable with aching heart.'[13] Tesla had found his vocation – and was desperately afraid that he would never be able to achieve it.

By now Tesla knew that he wanted to be an inventor. But he also knew that his father was determined that his remaining son should follow him into the church. Milutin's original ambition had been that his eldest son should become a priest, but following Dane's tragic death he had transferred those hopes onto Nikola. It was now his duty to fulfil the role that had been intended for his older brother and continue the family tradition. However repugnant Nikola found the thought of entering the priesthood – and it seems clear that he had no sense of calling at all to such a future for himself – he was also determined to be a dutiful son and to do what both his parents wanted. His dream of studying engineering and devoting his life to invention would have to remain just that – a dream.

Tesla's longing for a future as an inventor is revealing in itself about the world in which he was living. Regardless of his parents' ambitions for him, this was a world in which such a future seemed possible. It was a world where a career in invention was at least a plausible ambition, even for a young man living on the edges of the Austrian Empire. It shows that the nineteenth-century world really was getting smaller. Even in small towns in Croatia like Gospić and Karlovac, the tentacles of progress had penetrated. As those mechanical models and electrical demonstrations that Tesla encountered at school attest, science and technology were part of that penetration. Tesla already knew enough about physics to fantasize about an invention 'to convey letters and packages across the seas, thru a submarine tube', and even 'a ring around the equator which would, of course, float freely and could be arrested in its spinning motion by reactionary forces, thus enabling travel at a rate of about a thousand miles an hour, impracticable by rail'.[14]

Tesla got his way, of course. He became an inventor, not a priest. At the end of his final year of school at Karlovac he received a message from his father suggesting that he should take a hunting trip to the mountains rather than hurry home to Gospić. Puzzled, since hunting was not an activity usually favoured by his father, he disobeyed the instruction – and returned home to find the town in the middle of a cholera outbreak. Tesla promptly fell ill himself and was close to dying. His parents were distraught. With nothing to lose, perhaps, Nikola made one last bid for freedom. 'Perhaps I may get well if you let me study engineering,' he bargained with his father. Milutin's response was unequivocal: 'You will go to the best technical institution in the world.' As far as Tesla was concerned, it was as if 'a heavy weight was lifted from my mind.'[15]

Milutin insisted that his son spent a further year recuperating from his illness before setting off to continue his studies, but he kept his promise. Tesla was enrolled to study at the Joanneum Polytechnic School in Graz. His father had also secured for him a scholarship from the Military Frontier Administration Authority that would both fund his studies and allow him to defer his obligatory military service until he had completed his studies there. The Joanneum had been founded in 1811 by the Archduke John of Austria. In 1864 it became a Technische Hochschule, making it one of just four institutions in the Austrian Empire that offered degrees in engineering. The technical education that he would receive there would be among the best in Europe. The Technische Hochschulen had been established by the empire to provide precisely that – and to produce a generation of technically trained and proficient experts who would be at the forefront of the imperial drive towards its own technological future.

Chapter 2

Electric Power

In the summer of 1873, while Tesla was coming to the end of his schooldays at Karlovac and succumbing to cholera at home in Gospić, the world was coming to Vienna. The Vienna International Exhibition was officially proclaimed open on 1 May that year. It was, quite explicitly, about defining the empire's place in the new industrial world of the future. It was 'the great event that was to proclaim New Austria the peer in hopeful enterprise and self-improvement, of her elder sisters, England and France'.[1] Huge resources had been thrown at the event, and the sheer physical space occupied by the exhibition – 2,330,631 square metres – was enormous. The main building measured almost a kilometre from end to end. As an exhibition it was judged a huge success. As one commentator put it: 'No pencil will ever succeed in depicting its beauties, no tongue will give more than a feeble echo of its wonders.'[2] As a money-making exercise it was a complete failure, mainly because it coincided with a financial crash in the middle of the summer.

The Vienna exhibition was, of course, the latest in a line of industrial shows that had been inaugurated by the huge success of the Great Exhibition in London's Hyde Park in the summer of 1851. People had flocked to the Crystal Palace in their hundreds of thousands, with about 6 million visitors having passed through from the day the exhibition was ceremonially opened by Queen Victoria on 1 May until it closed its doors for the last time on 11 October. The Great Exhibition offered its visitors a frozen tableau of the past and present of invention, and invited them to imagine a future made up out of its contents. In a letter to her father, Charlotte Brontë called it 'a wonderful place – vast, strange, new and impossible to describe'. It was 'as if only magic could have gathered this mass of wealth from all the ends of the earth – as if none but supernatural hands could have arranged it thus, with such a blaze and contrast of colours and marvellous power of effect'.[3]

The exhibition in Hyde Park set the standard for the ones that followed. Just a few years later Paris organized its own Exposition Universelle, with a Palais de l'Industrie to rival London's Crystal Palace. Just over a decade after the Great Exhibition's runaway success, London hosted its second international exhibition in the summer of 1862. Electricity was an increasingly common presence at such shows. Visitors to the 1862 exhibition could marvel at the 'magnetic telltale of Professor Wheatstone' that 'telegraphically announced his or her arrival to the financial officers in whose rooms were fixed the instruments for receiving and recording the liberated current'.[4] There were electromagnetic dynamos too. At the next Paris Exposition Universelle in 1867 there was plenty of electricity as well. Jules Verne visited the exhibition while writing *Twenty Thousand Leagues Under the Sea* and some of the electrical technologies he encountered there found their way into his novel.

The visitors who flocked to the Vienna exhibition, just like those who had flocked to London and Paris, were there to find out what the future would be made out of. Among the exhibits they would find plenty of evidence to suggest that the shape of things to come would be electrical. There were exhibits of the latest telegraphic technology, for example. The German company Siemens & Halske had a variety of telegraphic apparatus on show. Visitors could inspect samples of undersea cables. As well as telegraphic apparatus, they exhibited a steam-driven electromagnetic generator in the Machine Hall and used it to power an electric light. In among the huge steam engines in the Machine Hall were a number of other electromagnetic generators as well.

One of the electrical exhibitors at Vienna was the Belgian inventor Zénobe Gramme. He was there to exhibit the new electromagnetic generator he had recently invented. The generator was notable in that it generated an almost entirely constant direct current, rather than the fluctuating alternating current usually produced. It was also extremely powerful. The story has it that at some stage during the exhibition, Gramme's partner Hippolyte Fontaine accidentally connected the Gramme generator to another one nearby, that happened to be generating current. He was amazed to see that the Gramme generator itself started moving. Not only was Gramme's device a generator, it was also a powerful engine. The story seems apocryphal. It seems more likely that Gramme and Fontaine were already aware of their device's potential as an engine, and that they chose the Vienna exhibition as a suitable opportunity to mount a public demonstration of their discovery. The Gramme machine, in any case, would have a significant role to play in Tesla's future.

By the time Gramme put his invention through its paces in Vienna, electromagnetic generators and engines had been around for almost half a century. From the 1820s onwards, electricians competed to find ways of making electricity useful. Hopes were high that electricity would power the future. As one commentator, the English electrician Alfred Smee, put it, 'to cross the seas, to traverse the roads, and to work machinery by galvanism, or rather electro-magnetism, will certainly, if executed, be the most noble achievement ever performed by man.'[5] Many electricians were convinced that electricity held the secret to understanding the universe, and that getting to grips with the stuff would not only offer 'a closer insight into the operations of nature as connected with the animal, vegetable or mineral kingdoms' but offer ways of 'usurping at no distant period, the place of steam as a mechanical agent', so that electricity would become 'in the most extensive manner, subservient to the uses, and under the control of man'.[6]

William Sturgeon's invention of the electromagnet in 1824 was the key to these early efforts to make electromagnetic engines. They generally worked by making use of the electromagnet's capacity to switch its magnetism on and off in rapid succession to produce either a reciprocating or a rotary motion. During the 1830s the American Thomas Davenport developed his own electromagnetic engines, making use of Joseph Henry's improved and more powerful electromagnets. With engines like these, technological optimists were confident that 'half a barrel of blue vitriol, and a hogshead or two of water, would send a ship from New York to Liverpool.'[7] Dreams of future electrical travel seemed on the verge of fulfilment in 1839 when the German-born Moritz Hermann von Jacobi sailed a boat with fourteen passengers on the river Neva in St Petersburg, powered by an engine working with a battery of the nitric acid cells just invented by the Welsh electrician William Robert Grove. In 1848 when the British Association for the Advancement of Science visited Grove's home town of Swansea, they were invited to see a boat powered by his nitric acid cells sailing around an ornamental lake at Penllergaer.[8]

In the United States the inventor and patent agent Charles Grafton Page succeeded in persuading a Congress usually averse to unnecessary expenditure to award him $40,000 to develop an electromagnetically powered locomotive. It took a decade to complete but on 29 April 1851 Page's 24-horsepower locomotive, powered by a bank of Grove's nitric acid cells, set off on a public demonstration trip from Washington DC to Baltimore. It was a triumph and a disaster. As Page's friend John Greenough reported, 'we carefully timed the revolutions of the driving wheels, and found that at our highest speed we had attained the unlooked for rate of nineteen miles an hour … propelled by some invisible giant, which by his silence was as impressive as his noisy predecessor, although less terrific.'[9] The speed was the problem though. As they were rattled around on the rails the fragile Grove cells started to shatter. The locomotive never made it to Baltimore, with Page calling an ignominious halt to the journey after only a few miles. It took two hours for the damaged locomotive to make it back to the starting point.

Electromagnetic engines like the ones that Page developed were just not powerful enough to be commercially successful – and banks of batteries were simply not a reliable or economic enough source of electricity. As Grove himself pointed out, it would only be when 'instead of employing manufactured products or educts, such as zinc and acids, we could realise as electricity the whole of the chemical force which is active in the combustion of cheap and abundant raw materials' that electricity would be an economic source of power. Only then would they 'have at our command a mechanical power in every respect superior in its applicability to the steam-engine'.[10] It was not really until the 1870s, when inventors like Gramme started to develop new kinds of electromagnetic engines, and new ways of generating electricity economically in large quantities, that the dream of electric power and locomotion started looking like a practical reality for more than a few visionaries like Page.

Locomotion was not the only potential use of future electricity, of course. The induction coil (another piece of electrical technology that, like Gramme's engine, would play a key role in Tesla's future) was another invention of the 1830s. Following the great experimenter Michael Faraday's observation of electromagnetic induction in 1831, a number of experimenters (including Faraday himself) tried to develop new instruments to show off the new effect. The first induction coils were made by the Irish philosopher-priest Nicholas Callan in 1836 and consisted of two coils of copper wire, one inside the other. When a current was switched on and off in the inner coil, a current was generated in the outer coil as well. It offered experimenters and lecturers a way of magnifying the spectacular effects of electricity in their demonstrations. As Callan put it, 'it supplies the place of all the various kinds of voltaic batteries, of the battery for producing a large quantity of electricity of low intensity, of the battery for exciting a large quantity of electricity of the intensity necessary for the rapid fusion and deflagration of metallic wires, and of the battery for producing an electric current of high intensity.'[11]

At the beginning of the 1850s the German instrument maker Heinrich Ruhmkorff, living in Paris, developed a technique for making induction coils far more powerful. Now they could become a source of serious spectacle. With Ruhmkorff coils it was possible to generate electricity of far higher intensity than could be generated from an ordinary coil. Another German instrument maker, Heinrich Geissler, found a way of exploiting this to spectacular effect with his invention of the eponymous Geissler tube. Geissler was a skilled glassblower and could produce his tubes in a variety of fantastical shapes. When they were filled with different gases and hooked up to a coil they glowed in different colours. Throughout the second half of the century, ever more extravagant Geissler tubes were essential items in the repertoire of popular scientific lecturers. They came in a variety of shapes and elaborate coils. Some featured fluorescent liquids as well as glowing gases, or were made from uranium glass that glowed green in the dark.

Natural philosophers used Ruhmkorff coils to probe the nature of matter, but there was plenty of scope for spectacle there as well. While investigating the phenomena of glowing gases in discharge tubes, the English experimenter John Peter Gassiot came up with one of the century's most spectacular experiments – the Gassiot cascade. In this experiment, a wine glass was placed inside an air-pump, standing on a brass plate connected to one pole of an induction coil; a wire connected to the other pole was suspended inside the glass. When the coil was connected and as the air-pump was evacuated of air, 'a discharge takes place in the form of an undivided continuous stream overlapping the vessel, as if the electric fluid was itself a material body running over. When first witnessed it appears at the moment impossible to divest the mind of such a conclusion.'[12] Gassiot carried out experiments with Geissler tubes too, 'in which many beautiful and novel results are produced'.[13]

The 'monster coil' unveiled by John Henry Pepper at the Royal Polytechnic Institution in London in 1869 was very much part of this culture of experiment as a source of wonder and sensation. The monster coil's dimensions – 150 miles of wire in the secondary coil and a primary coil that weighed 145 pounds – were a source of spectacle in themselves, even before the machine was put into action. The coil in action produced 'a spark, or rather a flash of lightning, 29 inches in length and apparently three-fourths of an inch in width, striking the disk terminal with a stunning shock'. When the terminals were brought closer, 'the discharge appears to issue more slowly as a gush of waving flame, and this flame may be blown away in a broad sheet, leaving the actual line of discharge unaffected and visible by its different colour.' It was 'a source of endless delight and wonder', said the breathless newspaper reports.[14]

People were fascinated by electric illumination. That was one of the attractions of Geissler tubes and Gassiot's cascade. Experimenters were already playing with the possibilities of electric light more generally. In 1845, for example, William Robert Grove invented an electric lamp he claimed was bright enough to read a newspaper by.[15] It is worth remembering from the perspective of a culture in which illumination is ubiquitous just how rare bright light was before the final decades of the nineteenth century. Most streets were dark; most houses were lit by candles or oil lamps. Even the gas lighting to be found in the houses of the more prosperous middle classes were dim by modern standards. Tesla, living in rural Croatia, would almost certainly never have seen bright light other than the sun while he was growing up.

In European cities, though, electrical illumination was starting to find its place by the middle of the century. Arc lights (very bright sparks between carbon electrodes), powered by electric batteries, started appearing around that time. An experimental arc light was installed on the Place de la Concorde in Paris in 1844. It was 'so strong that ladies opened up their umbrellas – not as a tribute to the inventors, but in order to protect themselves from the rays of this mysterious new sun'.[16] A few years later the inventor Edward Staite mounted a similar display on London's Trafalgar Square. The light 'produced the same sort of illumination as the sunlight through atoms of dust … Nelson's column, which was selected as the principal point, being frequently as conspicuous as at noonday'.[17] Similarly in 1855, Joseph Lacassagne and Rodolphe Thiers put their arc lamp to work in Lyon, where spectators 'suddenly found themselves bathed in a flood of light that was as bright as the sun'.[18] By the 1870s arc lights were relatively common. They were used in lighthouses, and were used to illuminate the city's fortifications by its defenders during the siege of Paris in 1870.

The connection between electric illumination and spectacle was clear in the widespread use of the arc lamp in theatrical productions. In May 1849 Edward Staite's electric light was even the star of its own show. The ballet *Electra*, mounted at Her Majesty's Theatre – and duly performed in front of Queen Victoria at a command performance a few weeks after its opening night – was commissioned

Edward Staite's arc light demonstration in Trafalgar Square.
(*Illustrated London News*, December 1848)

specifically to show off Staite's arc lamp. Electric illumination was becoming a staple of international exhibitions. At London's 1862 exhibition the crowds jostling to marvel at the display of electric light were so large the whole thing had to be moved. The following year electricity played a starring role in the annual London illuminations with St Paul's Cathedral decked out with electric lights.[19] Similarly, electric illumination played an important role in the Paris exhibitions of 1855 and 1867. Electric light featured at the Vienna exhibition both in its own right and as a way of showing off the power of Siemens & Halske's electromagnetic generators.

Electric light was starting to appear in domestic settings by the 1870s too, though Tesla in Gospić, Karlovac, or even Graz is unlikely to have encountered it at the time. In England, the wealthy arms manufacturer and enthusiast for electricity, W.H. Armstrong, was turning his country estate at Cragside into an electrical Xanadu. By 1871 he was running a Siemens generator on water power from a local reservoir to produce electricity to light up the house and grounds.[20] Arc lamps were little use for domestic, indoor purposes – the light they produced was simply too bright and penetrating. Battery power was of limited use in powering domestic lighting systems – batteries ran out too quickly and were difficult to deal with without at least some specialist knowledge. Electrical entrepreneurs were already dreaming about a future in which electricity would be generated in central power stations and carried into houses through networks of wires, just as domestic gas was distributed through a system of pipes.

Edward Bulwer-Lytton's *The Coming Race* (1871) offers a nice example of the ways in which electricity was starting to colour visions of the future, though again it's unlikely that the young Tesla encountered it. In Bulwer-Lytton's fantasy an American engineer encounters a subterranean race of superior beings, the vril-ya, whose culture revolves around a mysterious power they call vril, which readers would immediately have recognized as electricity. Making use of their power to control vril, the vril-ya have complete command of their environment. They can control the weather; they can even communicate telepathically. They have an array of powerful electrical weaponry at their disposal. This was what the future looked like to many at the beginning of the 1870s, as Tesla was beginning his studies at Graz. The 1873 Vienna International Exhibition was expressly designed to signal the Austrian Empire's desire to be a part of that future.

By the time of the next international exhibition in 1876 Tesla was already well advanced in his electrical education. Given his fascination with electrical invention it is difficult to believe that he was not following events at the Philadelphia Centennial Exhibition with obsessive interest. This was, after all, where the electrical future he wanted to be part of was in the process of being made real. The usual array of telegraphic apparatus was on show, including a printing telegraph that could operate without batteries, the electricity being produced by an electromagnetic generator worked by a treadle similar to the ones used in sewing machines. Thomas Alva Edison had some of his electric incandescent lamps on show. Gramme's company repeated the show they had given in Vienna a few years earlier, using one of their machines as a generator and the other as an engine working a pump.[21]

The eventual electrical star of the Philadelphia Centennial Exhibition was Alexander Graham Bell's telephone. The device itself did not look particularly impressive and did not initially receive much notice. After the two distinguished men of science Joseph Henry and William Thomson – one American, one Scottish – had pronounced the telephone 'the greatest marvel ever achieved in electrical science' it won one of the exhibition's coveted prize medals.[22] This was a real piece of the future on show, and Tesla at the Polytechnic School in Graz must surely have heard about it. It opened up new horizons for what electricity could achieve. Not just the dots and dashes of Morse code, but the human voice itself could be transmitted down the wire. Excited commentators were soon speculating that vision as well as sound could be transmitted electrically like this. As Tesla approached the end of his university education in engineering, the electrical future he was dreaming about seemed to be getting closer every day.

CHAPTER 3

Working Electricity

When William Fothergill Cooke and Charles Wheatstone acquired their patent for an electromagnetic telegraph on 10 June 1837, the world changed. It did not change immediately of course, but the possibilities opened up by the prospect of near-immediate communication between distant places had a profound impact on western culture. The cables that encompassed the globe by the end of the nineteenth century transformed the way that information travelled, and in the process changed the way that information was understood. Looking back at the early days of telegraphy, the English telegraph engineer Latimer Clark marvelled at the extent to which 'distance and time have been so changed to our imaginations, that the globe has been practically reduced in magnitude, and there can be no doubt that our conception of its dimensions is entirely different to that held by our fathers'.[1] Electricians had 'trained the electric agent as a dutiful child or obedient servant, to carry our messages through the air by the road we have made for it, and with equal velocity through the earth by a road it has made for itself'.[2]

The telegraph system that Cooke and Wheatstone patented was the direct product of contemporary fascination with scientific – and specifically electrical – spectacle. Electrical experiments were all about making the mysterious and invisible fluid visible in a variety of different ways. The trick of telegraphy was to find ways to make use of the spectacle to communicate information. One of the first telegraph receivers they developed illustrates this. It consisted of a diamond-shaped board marked with the letters of the alphabet and with a row of five magnetic needles along the middle. As the message was transmitted the needles converged to point at different letters, spelling it out. It was meant to be simple and easy to use – anyone who could read could use the telegraph, no special skill or knowledge was required. Making it all seem as simple as possible was essential in their efforts to find a market for their invention, and to convince railway entrepreneurs and engineers like Robert Stephenson and Isambard Kingdom Brunel that the telegraph had a role to play in managing the railways.[3]

Following the establishment of the Electric Telegraph Company in 1845 networks of telegraph lines started spreading out from London, following the growing railway network. These first telegraphs were mainly used for railway signalling purposes – to send messages down the line about delays and obstructions and to make sure that trains operated as efficiently as possible. They were often very simple devices – even simpler than the original Cooke and Wheatstone five-needle telegraphs. They were usually operated by railway station managers and needed little specialist knowledge. Maintaining the network was a different matter though. Those responsible for keeping the telegraph running, finding and repairing broken cables and so on, needed to know something about electricity. A new kind of specialist – the telegraph engineer – was needed. Charles Vincent Walker was an early example. He was an active electrical experimenter, having been one of the founders of the London Electrical Society and the editor of the *Electrical Magazine*. By 1845 he had succeeded in turning his electrical knowledge into a career, being appointed the superintendent of telegraphs to the South Eastern Railway.

People like Walker were instrumental in turning electricity into a profession. They came from a variety of backgrounds. Some, like

Walker himself, had established interests in electricity. Similarly, Cromwell Varley, who joined the Electric Telegraph Company as an engineer in 1846, came from a scientific background. Others, like the brothers Edwin and Latimer Clark, started their careers as civil engineers before turning to telegraphy. In 1861 a new journal, *The Electrician*, was established. Its editor, Desmond Fitzgerald, was himself a telegraph engineer and the new magazine was intended as a vehicle for the exchange of news and information between members of the new profession. A decade later in 1871 the Society of Telegraph Engineers was established. Its foundation signalled the coming of age of telegraph engineering. Its first president was Charles William Siemens, the younger brother of Werner Siemens, founder of the German firm Siemens & Halske. His presence in the new society's chair was a sign of the increasing internationalization of the telegraphic profession.

In the United States the artist Samuel Finley Breese Morse was the first to patent an electric telegraph. The inspiration came to him, he later claimed, while returning to New York from a long European tour on the packet ship *Sully* in 1832. During the voyage he found himself in conversation about the latest discoveries in electromagnetism with the Boston medical man Charles Jackson, during the course of which Morse himself suggested that if 'the presence of electricity can be made visible in any part of the circuit, I see no reason why intelligence may not be transmitted instantaneously by electricity'.[4] Back in New York he set about turning the sketch he had drawn on board the *Sully* into a working instrument. Morse acquired a patent for his projected instrument in 1837, a few months after Cooke and Wheatstone acquired their British one. After diligent lobbying, he also succeeded in persuading the US Congress to award him a grant of $30,000 to carry out experiments on long-distance electric telegraphy.

As Morse, like Cooke and Wheatstone in Britain, soon discovered, getting the telegraph to work needed a new sort of electrical knowledge. Electrical apparatus that performed perfectly well on a small scale in an electrician's laboratory was rather less reliable when deployed on a large scale. He needed help from partners who knew more than he did about batteries and coils. During 1837 he entered into partnership with Alfred Vail, who undertook to finance large-scale experiments on the transmission of electricity through long circuits, and he started to collaborate with Leonard Gale, professor of chemistry in New York, on those experiments. Among other things, Gale was familiar with the electrical researches of Joseph Henry on the best ways of arranging batteries to work powerful electromagnets. That was the kind of knowledge Morse needed to make his telegraph work over longer distances. By the middle of the 1840s he was in a position to turn his prototype into a real working telegraph, with the establishment of the Magnetic Telegraph Company.

Just as in Britain, the telegraph network in the United States expanded in conjunction with the railways. By 1861 an overland telegraph existed linking the east and west coasts of the continent. American telegraphy developed its own distinctive and very masculine culture. Being a telegraph operator was a job for young men. A small army of itinerant 'tramp telegraphers' were in high demand, particularly in the aftermath of the American Civil War. These men were highly skilled operators, adept at using Morse's code and apparatus and in constant competition with each other for speed and accuracy of transmission. They could travel the rails to wherever their skills were most needed and where the pay was best. The young Thomas Alva Edison was one of these itinerants. It was as a travelling telegrapher that he learned the basics of electricity and honed the skills of invention that would make him famous in later life.[5]

It seems likely that Morse would have come across accounts of early experiments in electric telegraphy during his European tour between 1829 and 1832, and may have seen some public demonstrations of telegraphy over short distances. The Russian diplomat Pavel Schilling had demonstrated the possibilities of electric telegraphy in St Petersburg in 1832. In 1836 the German astronomer and natural philosopher Carl August von Steinheil built a telegraph line linking the Bavarian Academy of Science in Munich to the Royal Observatory at Bogenhausen. In 1838 he discovered that the earth could replace the return wire in telegraphic circuits. It is clear that experiments like these had an important role to play in the development of British and American telegraphy. William Fothergill Cooke started thinking about telegraphy as a direct result of attending a demonstration of Schilling's telegraph, for example. But by the 1850s European states like the Austrian Empire were moving towards adopting variants of British and American telegraph systems.[6]

Unlike in Britain and the United States where telegraphy was largely promoted by commercial companies (after their initial grant to Morse of $30,000 the US Congress decided not to accept his offer to sell them the telegraph for a further $100,000), telegraphy in France was initially a state monopoly and reserved for government and military business until 1851. In the German states as well as in Tesla's homeland the Austrian Empire, national governments encouraged the adoption of the electric telegraph as part of their efforts to compete with Britain and France. On another European tour in 1845, anxious to convince European states and telegraph companies of the advantages of his own system, Morse devoted considerable efforts to trying to persuade Prince Klemens von Metternich, the Austrian Empire's powerful chancellor, to adopt it as the basis of the imperial telegraph system, for example. As European telegraph networks grew, the need for operators to run them and electricians to maintain them grew as well. These were appealing prospects for young men who wanted to be part of the future.

By the 1860s telegraphy was a major industry on both sides of the Atlantic. The activities of the Siemens & Halske company established in 1847 by Werner Siemens and Johann Georg Halske are a good example of how the telegraph industry worked, and how it expanded. Based originally in Berlin, the firm manufactured telegraph apparatus under licence from Cooke and Wheatstone. As well as manufacturing apparatus, they contracted to build and maintain telegraph lines for various European states. By the early 1850s Siemens & Halske had branches in England and in Russia, and had contracts to build and operate telegraph lines across Prussia and a number of other countries. In 1859 the company established a branch in Vienna, but the promised government contracts failed to appear, and it was closed down a few years later. By the 1860s telegraph networks were spreading well beyond Europe and North America. In 1870 Siemens & Halske completed the ambitious Indo-European telegraph line, connecting Britain to its empire in the East.

By the 1850s telegraph cables were being laid under the sea as well as over land. The first experimental submarine cable was laid across the English Channel in 1850. A year later a more permanent connection was laid between Dover and Calais. A year later in 1852 the Anglo-Irish Magnetic Telegraph Company laid down a cable under the Irish Sea connecting Holyhead in Wales with Howth, a few miles north of Dublin. In 1854 an undersea cable was laid linking Denmark and Sweden, and in the same year submarine cables were laid between Italy and the islands of Corsica and Sardinia. By the second half of the 1850s determined efforts were being made to lay submarine cables in deep water across the Mediterranean.

Telegraphy was in the process of becoming global. Speculators looked forward to a future in which the entire planet would be bound together in a network of telegraph cables.[7]

The first attempt to lay a telegraph cable under the Atlantic took place in 1858, though there had been speculation about the prospect since the 1840s. The Atlantic Telegraph Company was established in 1856 by the English cable manufacturer John Brett, the telegraph engineer Charles Bright, and the American telegraph entrepreneur Cyrus Field. The cable was manufactured by the Gutta Percha Company and consisted of six strands of copper wire wound around a central copper core and covered in three layers of gutta percha – an early variety of rubber used for insulation. The cable was laid down by HMS *Agamemnon*, on loan from the British government, and USS *Niagara* on loan from the United States. It was a huge technological undertaking, finally completed in August 1858 when congratulatory telegraph messages were passed back and forth between Queen Victoria and US President James Buchanan. The triumph was short-lived, however, since the cable soon stopped transmitting. It was not until 1866 that a permanent transatlantic telegraph link was finally established.

Following the eventual success of the Atlantic cable, further ambitious schemes for underwater telegraphy came to the fore. By 1870 British telegraph companies were laying cables along the Red Sea linking Suez to Aden, and across the Indian Ocean linking Aden to Bombay. In 1871 the British Indian Extension Telegraph Company laid a cable linking Madras to Penang, while the China Submarine Telegraph Company linked Singapore to Hong Kong. Submarine telegraphy was highly lucrative and competition between rival companies and nations was intense. Connections like these were important not just commercially, but strategically as well. By the end of the century maintaining an 'all red' telegraph network that linked the various parts of the empire (coloured red on contemporary maps of the world) without crossing any potentially hostile territory was a major British concern. British companies dominated submarine telegraphy. The early telegraphers' dream of conjuring up 'a spirit like Ariel to carry our thoughts with the speed of thought to the uttermost ends of the earth' had been realized.[8]

Realizing the dream had needed new knowledge, and a new sort of expert to wield that knowledge. In many ways the failure of the first Atlantic cable in 1858 was an important catalyst in the emergence of this new expertise. A huge amount of investment, not only in terms of finance but also in terms of national and individual prestige, had been poured into the project to cross the Atlantic telegraphically. Answers were needed as to what went wrong. Experiments by the Glasgow professor of natural philosophy and Atlantic Telegraph Company director William Thomson had already suggested that the cable used for the Atlantic cable was of variable quality and that different segments had different resistances. The Company's engineer, Wildman Whitehouse, insisted that resistance was irrelevant as long as they used his patent induction coil apparatus which could send rapid bursts of high-intensity electricity down the cable regardless.[9]

Whitehouse and his apparatus in the end turned out to be the villains of the piece. The increasingly high-intensity bursts transmitted down the cable as operators struggled to keep it working turned out to have been critical to its failure. The Atlantic Telegraph Company and the British government convened a committee to examine the causes of failure, and as part of the process commissioned new research on the electrical characteristics of telegraph cables. The telegraph engineer Latimer Clark carried out experiments on hundreds of miles of copper wire insulated with gutta percha. Another engineer, Fleeming Jenkin working at R.S. Newall's cable factory in Birkenhead near Liverpool, where some of the Atlantic cable had been made, carried out experiments to try to discover how much current would leak through the cable insulation and cause it to break down. It was the expertise of new men like these that would matter for the future of telegraphy and of electricity.

It was the Atlantic cable's initial failure (as well as its eventual success) that hammered home the message that the electrical future would need new ways of doing things. There was, for example, a pressing need for common electrical standards and units. Only by making sure that everyone manufacturing telegraph cables measured electrical resistance in the same way, using the same units and the same sorts of instruments could telegraph companies be sure that they were buying cables of the right quality. In Britain, following the Atlantic cable fiasco, the British Association for the Advancement of Science established a committee on electrical standards to deal with the problem. The British attempt to establish a standard unit of electrical resistance – the ohm – was centred on Cambridge and the new Cavendish laboratory. In Germany, the pressure came from Werner von Siemens, who put up the money and persuaded a sceptical Otto von Bismarck that the new Germany needed its own Physikalisch-Technische Reichsanstalt to ensure its place in the electrical future.[10]

By the time Berlin's PTR was eventually established in 1887, Tesla was already an accomplished electrical engineer. His studies at Graz would have provided him with the basic understanding of physics that he would need to survive and prosper in this brave new electrical world. He attended lectures on physics by Jacob Pöschl – a 'methodical and thoroughly grounded German', as Tesla described him, whose 'experiments were skilfully performed with clock-like precision and without a miss'. During the course of those lectures and demonstrations he would certainly have encountered the rudiments of electricity. He would have learned about Faraday's experiments, about different sorts of batteries, and about the electric telegraph. During Tesla's second year at Graz, Pöschl acquired one of the new Gramme dynamos (as demonstrated at the Vienna Exhibition in 1873) and showed it off to his students. It was Tesla's first real encounter with this new technology, and it fired his ambition to become master of it.[11]

Tesla left Graz without finishing his studies, however. He fell ill, probably as a result of too much intensive studying, and failed to complete his final year. He was disillusioned by his father's apparent dismissal of his hard work as a student, and had spent that last year gambling and drinking instead of attending his lectures. He left Graz and found a job in the town of Maribor in Styria to finance his activities. Even a despairing visit from his father failed to persuade him to resume his studies. He was eventually arrested as a vagrant and forcibly deported back home to Gospić. Shortly afterwards, his father died, causing further familial and personal turmoil. After his death, Tesla found a cache of letters written to his father by some of his professors at Graz, warning him that his son was in danger of killing himself from overwork.[12]

Tesla eventually moved to Prague and enrolled at the Karl-Ferdinand University there, taking lectures in various scientific subjects. He failed to complete his studies at Prague as well, and eventually gravitated to Budapest, hoping to find work in connection with plans to build a telephone exchange there. Those plans fell through, but instead Tesla found work at the Hungarian Central Telegraph Office. It was an opportunity for him to become familiar with the practical realities of telegraphic and electrical work. He found himself working hard at 'calculations, designs and estimates', and acknowledged that the 'knowledge and practical experience I

gained in the course of this work was most valuable and the employ-
ment gave me ample opportunities for the exercise of my inventive
faculties'.[13]

Tesla's background had given him an enviable combination of
theoretical and practical know-how. His studies at Graz and Prague
had given him his familiarity with the latest theories of electricity as
well as the mathematical skills to work with them. The hard grind
and routine of the Central Telegraph Office at Budapest had given
him essential experience of the realities of doing electrical work in
the world outside the laboratory or the lecture theatre. It was not
long before it paid off. In 1882, Ferenc and Tivadar Puskás – the two
brothers whose plans to build a telephone exchange had attracted
him to Budapest in the first place – invited him to Paris to work for
the Société Electrique Edison, established there to manufacture and
install Thomas Alva Edison's incandescent lighting system. It was
another key step in Tesla's journey into the future. In the 45 years
since Cooke and Wheatstone's patent for the electromagnetic telegraph had inaugurated the new electrical world, that world had
changed beyond recognition. It needed a new kind of electrician to
operate in this new environment, and Tesla was one of them.

In Paris, Tesla had plenty of opportunities to hone his skills.
He was working – if at a distance – for Edison himself, who by
the beginning of the 1880s had made a reputation for himself as
the greatest living electrician. As we shall see later, Edison had
worked hard at creating this image of himself. He recognized, as
Tesla himself would recognize in due course, the importance of
telling the right kind of story about himself and his origins as an
inventor. In 1881 Edison's display of his latest electrical products
at the Paris International Electrical Exhibition had been a huge
success and had helped his companies establish a secure toehold in
the lucrative European marketplace. Charles Batchelor, who had

overseen the Edison exhibit at the Exhibition, stayed on in Paris to manage Edison's companies there – and it was Batchelor who would be responsible for Tesla's fate over the following two years.

Batchelor certainly recognized that Tesla had that rare combination of skills in both the theory and the practice of electrical engineering. He kept the young employee busy. Tesla was soon developing new dynamo designs for the incandescent light systems that were being manufactured at the Société Electrique Edison's factory at Ivry, just outside Paris. Before long, he was being sent around the country to deal with technical problems that arose in the Edison lighting power stations that were being set up as the company expanded its markets. He spent several months at the end of 1883 in Strasbourg in Alsace (German territory in the aftermath of the Franco-Prussian War a decade earlier) dealing with a particularly intractable problem there. But even working for Edison in Paris and Strasbourg, Tesla was looking to the future – and in that future, America beckoned.

PART 2

BATTLE OF THE SYSTEMS

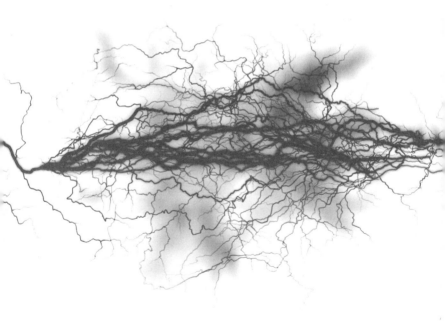

CHAPTER 4

A New World

When Tesla arrived in America he was entering a new world that seemed to be both primitive and modern at the same time. Recalling the event decades later, he remembered it as a sort of Arabian Nights journey in reverse. Rather than transporting him to paradise, the 'genii had carried me from the land of dreams into one of realities'. The Europe he had left behind him 'was beautiful, artistic and fascinating in every way; what I saw here was machined, rough and unattractive'. He felt that he had arrived in a country that was a century behind the Old World in everything that mattered. It did not take him long to change his mind, and he later said that when travelling abroad in 1889, five years after his arrival in America, 'I became convinced that *it was more than one hundred years AHEAD of Europe* and nothing has happened to this day to change my opinion.'[1]

The particular 'genie' that brought Tesla to America was, of course, Thomas Alva Edison. Tesla had been working for Edison's companies in Paris for two years by the time he received the offer to work for him in New York at the Edison Machine Works. The Machine Works (like its equivalent in Paris where Tesla had previously been employed) was a company established to turn the dream of an electric future into a reality. The urgent electrification that was taking place across Europe and America during the 1880s and 1890s needed infrastructure. Someone had to build the dynamos that would generate electricity for lighting and power systems. Someone had to install them and repair them when they broke down. This was technology at the edge of tomorrow, and the combination of theoretical knowledge and hands-on skill needed to deal with the new technology's idiosyncrasies was rare. That made Tesla himself a very desirable commodity too – hence his voyage to New York and the future.

The United States of America during the 1880s was a country that was still busily reshaping itself in the aftermath of a bloody and divisive civil war. That war had been fought, in part, over the question of American identity, and the American future. There was a growing sense – and not only in America – that the future would be a very American place. Getting there would require a state of mind and a way of thinking that many Americans felt belonged particularly to them. To a young man like Tesla, the United States seemed to offer an opportunity to remake himself as well. It was a country that sold itself to the world as the place where destiny was in the hands of the individual. It was a country that depended on and prided itself on ingenuity and invention. This was the place where fortunes and futures alike were waiting to be made.

The tradition of inventiveness that Tesla admired already had deep roots when he arrived in New York, and New York was one of the places where the tradition was most deeply embedded. From 1825, the city was home to one of the outposts of Charles Willson Peale's Museum, originally established in Philadelphia.[2] The New York museum had been established by Peale's son Rubens and featured the same mix of natural history, art, and spectacular invention as the original. That original Philadelphia museum was probably the inspiration for the National Gallery of Practical Science in London as well – established by the Philadelphian inventor Jacob Perkins in 1832.[3] In 1841 Rubens's New York museum acquired a powerful rival in the form of P.T. Barnum's American Museum, which offered a similar repertoire of exhibits.[4] The city's fascination with the future was exploited by the *New York Sun* when it hired Edgar Allan Poe to fool their readers with an account of the first balloon flight across the Atlantic in 1844.[5]

Places like Peale's or Barnum's museums helped forge a link between science, invention and spectacle that would be critical to the way Americans imagined the future for the rest of the century. Not all American men of science approved of this linkage of science and spectacle. Joseph Henry at the Smithsonian Institution, established in 1846, had a different vision of how to get to the American future. He and his allies (they called themselves the Lazzaroni) advocated an approach to science that was collective, systematic, and at the service of the American state. This was a vision of science as a disciplined and sober vocation. American men of science, argued the Lazzaroni, could only compete with their European counterparts if they partook in the same sort of systematic process of training. They – and the knowledge they produced – were meant to be the products of rigorous mental discipline, not flashy, self-made showmen.[6]

Despite the Lazzaroni's best efforts, the image of the man of science as a self-made, showy inventor was a very durable one. Ironically, Joseph Henry himself could be portrayed in such a way. He had first made a scientific name for himself while teaching at the Albany Academy with his construction of a series of increasingly powerful electromagnets. Before teaching himself science he had toyed with the prospect of a career in the theatre – his was a classic example of the scientific rags-to-riches life so admired by many of his fellow Americans. Another example was the Vermont blacksmith Thomas Davenport, who taught himself practical electricity after seeing one of Henry's electromagnets, and who in 1837 took out a patent for an electromagnetic motor, putting his invention on show in exhibition halls on both sides of the Atlantic.[7] Samuel

Morse was yet another inventor keen to promote an image of himself as self-made in science.

The magazine *Scientific American* was first published in New York in 1845, its establishment roughly coinciding with the founding of the Smithsonian Institution. But where the Smithsonian under Joseph Henry's direction soon came to embody the Lazzaroni's ambitions for organized and expert science, *Scientific American* was to be the champion of science by and for the practical inventor. These were contrasting visions of how America should get to the future. The magazine's founder, Rufus Porter, was himself a prolific inventor. He promised to provide his readers with 'general notices of the progress of Mechanical and other *Scientific Improvements*; American and Foreign Improvements and Inventions; Catalogues of American Patents; Scientific Essays, illustrative of the sciences of Mechanics, Chemistry and Architecture; useful information and instruction in various Arts and Trades; Curious Philosophical Experiments; Miscellaneous Intelligence, Music and Poetry'. Porter promised that his publication was going to be devoted to the interests of 'Mechanics and Manufacturers'.[8]

The first issue of the new publication opened with a description of 'improved rail-road cars' that detailed 'a variety of excellent improvements in the construction of trucks, springs, and connections, which are calculated to avoid atmospheric resistance, secure safety and convenience, and contribute ease and comfort to passengers, while flying at the rate of 30 or 40 miles per hour'.[9] Readers were told about Professor Faraday's researches on the properties of zinc, the latest developments in electroplating, and a new kind of daguerreotype camera. They were promised an early account of Morse's amazing telegraph. Subsequent issues carried accounts of everything from the new electric light to the latest labour-saving agricultural implements. These were the ingredients of the future, laid out for the magazine's readers, and Porter and his successor as editor were clear that this future would be the product of the inventive practical man.

For the next several decades the *Scientific American* continued to sell its readers this vision of an American future built out of the inventive genius and practical knowhow of the 'intelligent and liberal' common man. In 1855 the magazine approvingly quoted a lecture by James T. Brady: 'To the genius, talent, and industry, which mechanically apply the powers of nature in developing her resources, and the achievement of useful mechanical results, we may confidently look for the distinctive superiority of our people. Excellence in contributing toward this reputation should be esteemed second to none. And we should learn to think lightly of the mind or heart of him who would not cheerfully turn away from the exploits of Caesar, Hannibal, or Napoleon, to dwell with joy and emulation over the triumphs and the fame of Fulton, Whitney, and Morse.'[10] This was a vision of the future that made invention a peculiarly American property.

For those who subscribed to the *Scientific American*'s views of the American future, the Centennial Exhibition held in Philadelphia in 1876 to celebrate the centenary of the declaration of independence was an opportunity to show just how far along the road to the future America already was. Where the Great Exhibition of 1851 and the subsequent European exhibitions had demonstrated the Old World's progress towards the future, the Philadelphia exposition was going to be an American affair that would put the Old World in its place. The main Exhibition Hall, designed by the architect Henry Pettit and built by the engineer Joseph Wilson, would be the largest building ever built, covering more than 21 acres. It cost a staggering $1,580,000 to build. Like the Great Exhibition's Crystal Palace, the Exhibition Hall was meant to be a powerful symbol of how the future would appear.

The highlight of the opening day was the turning on by Ulysses S. Grant, the President of the United States, and Dom Pedro, Emperor of Brazil, of the great Corliss steam engine that dominated the main building. As the *Scientific American*'s correspondent put it, it 'was a scene to be remembered; and perhaps for the first time in the history of mankind, two of the greatest rulers in the world obeyed the order of an inventor citizen'.[11] The gargantuan engine, 45 feet in height and with a flywheel 30 feet in diameter, was an emblem of American power. More practically, through a series of shafts more than a mile in total length, it powered almost all the machines in the exhibition. Americans flocked to the Philadelphia exhibition in their hundreds of thousands. Those who could not come in person could witness it at a distance in the pages of the *Scientific American*, which started publishing a weekly supplement to cover the occasion.

'That the Centennial, both intrinsically as a display and in the circumstances connected with it, has been successful far beyond the lot of all previous world's fairs, is plainly evident,' thought the *Scientific American* at the end of the affair. The exhibition had been a success far beyond the most optimistic anticipations. It had shown Americans themselves, as well as their Old World rivals, just what they could achieve: 'Eleven years ago, these eight million people were engaged in a bitter and terrible internecine war. Now, great national gatherings have taken place day after day, unmarred by a word of sectional strife or ill feeling. For three years the nation has been suffering under a shrinkage of values and a financial stress which has brought ruin to thousands, and of which no one has escaped the evil effects. Yet despite all the privations and suffering incident thereto, a vast national enterprise has not only been successfully carried through, but has included such a representation of the fruits of American industry and genius as has never before been seen.' The country had 'learned to compare our own work with that done in Europe; and having found where we are excelled as well as where we excel, we have stored up a stock of ideas, sure to bear rich fruit in the future'.[12]

America was starting to look like the future's country – and not only in America. When Edward Bulwer-Lytton published what was to be his last novel – *The Coming Race* – in 1871 (a few years before the Philadelphia exhibition) it was no surprise that the story's human protagonist and narrator was an American 'with a taste for travel and adventure'.[13]

Scientific romances, tall tales and speculations about the shape of the future occupied an increasingly important place in the popular magazines that played such a vital role in American public culture during the final quarter or so of the nineteenth century. Magazines like *Harper's Weekly*, *Pearson's* and *The Century Magazine* offered their readers new ways of engaging with contemporary American culture and developing different ways of imagining the future was central to their appeal. Authors like William Livingston Alden, Cleveland Moffet, and even Mark Twain moved effortlessly between fact and fiction as they wrote for the magazines. Stories like Alden's *Van Wagener's Ways* might poke fun at inventors and their pretensions but the satire was firmly grounded in the assumption shared by authors and their readers that the future really was going to be transformed by technology.[14]

Tales of spectacular invention sold newspapers. In 1877 the same *New York Sun* that 33 years earlier had published Edgar Allan Poe's balloon hoax regaled its readers with an account of a new electrical device for seeing at a distance. On 29 March they published a letter declaring that an 'eminent scientist of this city, whose name is withheld for the present, is said to be on the point of publishing a series of important discoveries, and exhibiting an instrument invented by him, by means of which objects or persons standing or moving in any part of the world may be instantaneously seen anywhere and by anybody'. The fabulous device would 'supersede in a very short time the ordinary methods of telegraphic and telephonic communication'. Through it, 'mothers, husbands and lovers' would be able to see their loved ones; scholars could 'consult in their own rooms any rare and valuable work or manuscript in the British Museum, Louvre, or Vatican, by simply requesting the librarians to place the book, opened at the desired page, into this marvellous apparatus'.[15]

The *New York Sun* was exploiting the public's fascination with Alexander Graham Bell's new telephone, unveiled at the Philadelphia exposition just a few months earlier. Just as the telephone looked like the future of telegraphy, so the new distance viewer looked like the future of the telephone. Reports that such a device – usually called the telectroscope – was on the verge of being made public were repeated several times over the next few decades. In 1880, for example, the inventor George Carey claimed in the pages of *Scientific American* that the 'art of transmitting images by means of electric currents is now in about the same state of advancement that the art of transmitting speech by telephone had attained in 1876'.[16] No such device was ever actually made, but to Americans fixated on the possibilities of the future its reality seemed entirely plausible.

Some people thought that the scale and speed with which America was transforming itself had the potential to overwhelm its population. In 1881 the neurologist and authority on the medical uses of electricity George Miller Beard published *American Nervousness*. In it, Beard argued that a 'new crop of diseases has sprung up in America, of which Great Britain until lately knew nothing, or but little. A class of functional diseases of the nervous system, now beginning to be known everywhere in civilization, seem to have first taken root under an American sky, whence their seed is being distributed.'[17] This new crop of nervous disorders were uniquely American: 'no age, no country, and no form of civilization, not Greece, nor Rome, nor Spain, nor the Netherlands, in the days of their glory, possessed such maladies.' They were the direct result of the breakneck speed with which Americans were rushing towards the future.

Americans were like electric batteries that did not produce quite enough power for the work they needed to do. They did not generate quite as much nervous power as the pace of modern life required. Their bodies could not keep up with the future: 'while modern nervousness is not peculiar to America, yet there are special expressions of this nervousness that are found here only; and the relative quantity of nervousness and of nervous diseases that spring out of nervousness, are far greater here than in any other nation of history, and it has a special quality. American nervousness, like American invention or agriculture, is at once peculiar and pre-eminent.'[18] In fact electricity was one of the remedies touted for American nervousness, and Beard was himself an advocate of medical electricity. Those who had suffered breakdowns through rushing into the future at too headlong a pace might also try hydrotherapy, phototherapy, or calisthenics to restore themselves to a proper equilibrium.

The image America increasingly projected of itself during the final decades of the nineteenth century was of an industrializing, nervous and progressive country. It was a country in the process of transforming itself at an astounding pace. This was the future-embracing country that Tesla encountered when he disembarked from the *City of Richmond* on 6 June 1884. New York was a city that prided itself as a place where fortunes could be made. Tesla himself was certainly determined to make his fortune there. The New

World offered him opportunities to make himself in new ways, and Tesla was ready to embrace the future that America offered. His training as an electrical engineer in Graz and in Prague, as well as the practical experience of working with electrical machines that he had gained in Budapest and Paris, had given him the tools that he needed. He was going to make himself rich by putting himself to work in inventing the future. First, however, Tesla would have to deal with Edison.

CHAPTER 5

The Wizard of
Menlo Park

Thomas Alva Edison was the reason that Nikola Tesla went to America. If there was a single human being in 1884 that Tesla wanted to emulate, Edison was that man. Edison epitomized invention and the future in the 1880s. Tesla remembered his first encounter with Edison as 'a memorable event in my life'. He 'was amazed at this wonderful man who, without early advantages and training, had accomplished so much'.[1] The infatuation underlines the extent to which Tesla was already in thrall to the image of invention that Edison had made his own. Edison was an entirely self-made man – and that is how Tesla wanted to be as well. This image of invention was a powerful and seductive one. It clearly appealed to Tesla – as it appealed to many others at the end of the nineteenth century. When success came his way, Tesla would himself be celebrated as the epitome of the self-made inventor of the future.

Edison's view of Tesla on their first encounter was rather more prosaic. Their first actual meeting took place a few days after Tesla's arrival in New York. The dynamos on the SS *Oregon* had been damaged and Tesla, as the new boy, had been given the job of repairing them. Having spent the night working, he was returning home when he happened to meet Edison, himself out for a morning stroll. Having discovered that their 'Parisian' had already completed the task, Edison remarked (in Tesla's hearing) that he was 'a damn good man'.[2] Edison may have been impressed, but it's also clear that as far as he was concerned Tesla was simply another one of his cadre of skilled electrical engineers. He might be unusually keen, but he was there to work according to Edison's and his managers' instructions rather than to develop an independent life of invention of his own: there was to be only one Thomas Alva Edison.

A cartoon published in the *Daily Graphic* for 9 July 1879 captures very well indeed the way Edison looked (and wanted to look) to the world at the beginning of the 1880s. The cartoon, on the front page of the newspaper, was titled 'The Wizard's Search' and portrayed Edison dressed in a wizard's robe and carrying a blazing light in one hand as he apparently searched through a cave. His robes were embroidered with illustrations of his various inventions (with the recently invented phonograph particularly prominent) and his tall wizard's hat was decorated with a picture of a galvanometer. The search referred to by the caption was the quest to find a suitable material for the filament of the incandescent electric light that Edison's Menlo Park laboratory was conducting, and referred in particular to a recent appeal Edison had made for sources of platinum. It was the *Daily Graphic*, a few years earlier, that had first coined the epithet of 'Wizard of Menlo Park' for the inventor.[3]

That article in the *Daily Graphic*, written by the journalist William Augustus Croffut, was one of a number of similar celebrations of Edison that appeared around this period, building in particular on the reputation that he acquired in the aftermath of his invention of the phonograph. These newspaper articles, with Edison's own enthusiastic connivance, were instrumental in establishing him as a very particular kind of individual. Edison as portrayed in the popular press was very much a self-made man. While this kind of image of the inventor was hardly new in the 1870s – there was already a long history of celebrating the self-made and self-reliant ingenious individual on both sides of the

Thomas Alva Edison dressed as the 'Wizard of Menlo Park', with his robes decorated with his inventions. (*Daily Graphic*, 9 July 1879)

Atlantic by then – it reached new heights with Edison's portrayal. His inventiveness was the product of his own unique capabilities and idiosyncrasies. Even the ways he slept and dressed were portrayed as part of the pattern of idiosyncratic individuality that gave him an edge over his competitors.

Edison – or rather the public portrayal of Edison the inventor – taught Tesla how to be an inventor himself. Whatever their differences as individuals and competitors later in Tesla's career, Edison was an important example of what a life of invention might look like. The effort that Edison devoted to making himself as an inventor taught Tesla that there was more to the business of invention than creative skill, technical knowledge and imaginative ingenuity. The making of an inventor required spectacle and public performance as well. Edison excelled at selling himself in this way. He was a consummate showman and a dedicated networker. These were skills that Tesla would need to acquire as well if he wanted to succeed as an inventor. Edison was particularly adept at selling his own history as a rags-to-riches success story. It was a story that drew on well-established patterns of just what such a life of self-improvement should look like.

Born in Milan, Ohio in 1847, Edison was well placed to be presented to the world as a son of enterprising and industrious America. Milan at the end of the 1840s was booming. Positioned as the town was on the deepwater Milan canal connecting to Lake Erie, it was an important conduit for trade between the Great Lakes region and the east coast. Agricultural products flowed one way to feed cities like New York, and manufactured goods from those cities flowed the other. Edison's own family were relatively recent arrivals, drawn to the town by its growing prosperity and the opportunities it offered. A few years after Edison's birth, as Milan's fortunes declined with the arrival of the railways, the family followed the money to Port Huron, Michigan with its thriving lumber industry. There their fortunes fluctuated, with Edison's father Samuel trying his hand at a variety of businesses as well as farming ten acres of land.[4]

Just as would be the case with Tesla a few decades later, Edison and his biographers could point to a history of precocious ingenuity

that prepared him for a future of invention. Aged about twelve he
built a telegraph line a mile and a half long connecting his parents'
house to his friend James Clancy's home. It was an exercise in home-
spun ingenuity: the 'insulators were bottles set on nails driven in trees'
and the 'magnet wire was wound with rags for insulation'.[5] In a letter
to Edison in 1878 Clancy reminded Edison how they 'had a office
underground in your Fathers' garden and one in my house'. The same
friend recalled Edison's enthusiastic performance of chemical experiments that led his mother to warn them that they 'would yet blow
our heads off'. Others remembered him engrossed in mechanics'
journals, or building cannon, steam engines and a water mill.[6] These
were stories of inventive genius in the making. Each time they were
retold they reinforced the myth of Edison's precocity and the sense
that his passion for making things was somehow peculiarly his own.

By the time he was in his mid-teens Edison was working on
the Grand Trunk Railway between Port Huron and Detroit as a
newsboy. He was soon taking advantage of life on the railways and
the opportunities it offered, opening his own newspaper stand and a
vegetable stand in Port Huron, transporting his goods in the train's
mail carriage free of charge. It was not long before he was publishing his own newspaper. The railways taught Edison the value
and possibilities of telegraphy as well. He taught himself Morse
code – and mastered the basics of commercial operation. By the
time he was sixteen he was working as a telegraph operator on
the Grand Trunk line. Accounts of his precocious childhood walked
a fine line between showing him off as a typical example of ingenious American entrepreneurialism on the one hand and as uniquely
gifted on the other. He both embodied American values of gung-ho
individualism and went beyond them.

Life as a telegraph operator on the railways during the 1860s
certainly gave Edison plenty of opportunities to study the workings

of the communication network that was coming to symbolize the future. The telegraph network was expanding rapidly and there was a shortage of technically proficient young men like Edison to keep it working. Over the next few years he travelled widely, honing his skills and his technical knowledge. Anecdotes about his skill and speed as an operator – arriving in Boston and pushing his New York counterpart to the limits of his transmitting capacity in a test of his endurance – jostled with accounts of a serious-minded young student with his nose in the pages of the *Telegrapher* journal, reading about the latest developments in the field. This was already a young man with an eye for opportunity. On one occasion he persuaded another agent to let him have the electrodes from a pile of old, disused Grove cells. The electrodes were of platinum (the agent thought they were tin) and Edison could boast to his biographer that 'the reworked scraps are used to this day in my laboratory over forty years later.'[7]

Edison's first patented inventions came straight out of his experiences as a jobbing travelling telegrapher. As he worked the lines he scribbled potential inventions in his notebooks, including a variety of designs for duplex or diplex telegraphy (sending more than one signal through a line simultaneously, either in the same or opposite directions), and a system of secret military signalling. This was where Edison's experiences as a working telegrapher offered him a real advantage. He was familiar with telegraphy's limitations and its possibilities. His first patent, in 1869, was for an electric vote recorder, and was based on the electrochemical recording technology used in automatic telegraphs. This first patent was followed by an improved printing telegraph and a fire alarm telegraph. He also set himself up as a provider of private telegraph lines and established a telegraphic stock quotation service.

At this stage in his career Edison was one of many experienced telegraph men looking for inventive opportunity. Success depended

on his ability to extrapolate the future and imagine new uses for telegraphic technologies. It also depended on Edison's ability to attract
money. Patenting in America was a relatively expensive business.
Testing new devices and new systems needed time and resources.
Potential customers had to be persuaded that investing in Edison's
inventions would offer them a competitive edge. One of the lessons that Tesla would have learned by contemplating his hero was
that imagination was not enough. A life of invention needed to be
grounded in practical reality. It was the need for financial backing –
the need to be in the right place and in proximity to money – that
determined Edison's move to New York in 1869, just as Tesla in turn
would gravitate there a decade and a half later.

New York looked to Edison like the future's city – as it would
look to Tesla – because it was a city that already combined many
of the ingredients that would be needed to make the future real.
For the next decade or so, the city was Edison's base as he went
about the business of invention. He had 'entered definitely upon
that career as an inventor which has left so deep an imprint on
the records of the United States Patent Office'.[8] He developed a
quadruplex telegraph in 1874 that allowed four separate signals to
be transmitted simultaneously down a single wire. In the aftermath
of Alexander Graham Bell's invention of the telephone in 1876 he
developed his own improvements to the technology. But Edison
at this stage of his career was primarily an inventor's inventor. He
was known to his fellow telegraph engineers rather than being a
major figure in the public eye. That would change in 1877 with his
invention of the phonograph.

The invention of the phonograph, said Edison's biographer,
'was the result of pure reason'.[9] Edison himself described it as the
outcome of his attempts to find ways of automatically recording
telegraph messages. But the sensation that the new invention caused

was very much the product of Edison's adroit handling of the press. Phonograph in hand, he went to see the editor of *Scientific American* and told him he had 'a machine that could record and reproduce the human voice'. Famously, he recited 'Mary had a little lamb' and played the recording. As Edison put it, they 'kept me at it until the crowd got so great Mr. Beach was afraid the floor would collapse; and we were compelled to stop'. The following day's newspapers were full of reports of his spectacular invention and performance.[10] It was a piece of showmanship that would certainly have its impact on Tesla in due course. It would have taught him that spectacle and successful invention were inextricable.

Performances like Edison's with the phonograph were central to the process of making invention look like an act of exceptional individualism. Behind the scenes, however, Edison was already in the process of making invention collective. At the end of March 1876 Edison transferred his operations to a new laboratory he had built at Menlo Park in New Jersey. There he developed a new way of managing invention. Menlo Park was going to be an 'invention factory' designed to churn out patentable inventions and improvements on a regular basis. It was going to take the element of chance out of the process. The building's ground floor contained an office, a library, a model room and a 'machine shop, completely equipped, and run with a ten-horse-power engine'. Upstairs was the laboratory itself: 'Scattered through the room are tables covered with electrical instruments, telephones, phonographs, microscopes, spectroscopes, etc. In the centre of the room is a rack full of galvanic batteries.'[11]

It was the resources of Menlo Park and the people who worked there that made it possible for Edison to develop his electric light and power systems in the way that he did. Gushing newspaper articles might celebrate the Wizard of Menlo Park as an individual iconoclast but the work that took place in the laboratory there in reality was collective. It might be Edison's own name on the patents that flowed out of the building on a regular basis but the work that lay behind those inventions was carried out by a small army of assistants. It was that small army, for example, that carried out the extensive and laborious testing of different materials that led to Edison's invention of the incandescent electric light bulb. It was the dependable regularity with which inventions flowed out of the lab that made Edison an increasingly promising investment for potential financiers as well. Such people wanted a reliable return for their money and Edison's factory production line of invention all but guaranteed it.

So by the time Tesla encountered his hero in the flesh for the first time a few days after his arrival in New York in 1884, Edison's process of invention was very well worked out – and it depended on a steady supply of young men like Tesla. Edison himself was the very public face of an increasingly complex web of commercial organizations and agreements. As well as Edison's laboratory – about to move from Menlo Park to West Orange – his enterprises were divided into a number of commercial manufacturing and supply companies both in Europe and in the United States. These included the Société Electrique Edison, where Tesla had worked in Paris, and the Edison Machine Works where he started work in New York on 8 June 1884. Tesla and Edison were at different ends of the process of inventing the future. Being one of Edison's backroom boys certainly gave Tesla plenty of opportunities to discover how the pragmatics of commercial innovation really worked and what the business of invention really entailed.

It is impossible to know exactly what Tesla expected of Edison at this stage in his progress. He had certainly come to New York with high ambitions and an equally high opinion of himself and his capacities. He had his own compelling vision of how the future might be made. He also had some very concrete and specific ideas about how to make that future a reality. Tesla had been thinking about alternating current dynamos and motors ever since he had first seen a Gramme dynamo in Jacob Pöschl's classroom in Graz. He was convinced that this was where the future of electric power lay. Working for Edison in New York offered him his first real opportunity to turn the ideas he had been developing in his head into a practical technology. For the first time he had access to the materials and the tools he would need to build his vision. In Edison, he might also have imagined he had a potential financial backer for his schemes.

But it would not have taken Tesla long to realize that Edison was already committed to his own vision about the future of electrical generation, transmission and use. He was unlikely to support Tesla's alternative one, and Tesla in the end was cautious enough to keep his ideas to himself. In the meantime, Edison had set the 'Parisian' the task of developing a system of electric arc lights for public illumination, since Edison's own incandescent light system was thought unsuitable for street lighting. Tesla proceeded to carry out his task and expected to be rewarded for his diligence. In the meantime, however, Edison had entered into an agreement with another company to install their arc light systems for public street lighting. As a result, Tesla's system was never adopted, and he did not receive the financial reward he thought he had been promised. Disgusted by his treatment – and barely six months after starting to work for Edison in New York – Tesla resigned his post. He had learned the hard way the harsh realities of the business of invention.

CHAPTER 6

AC/DC

The years that followed Tesla's departure from the ranks of Edison's employees were often difficult, as we shall see. They culminated in triumph, however, and it is with that triumph that we begin this chapter.

On 16 May 1888 Tesla delivered a lecture to the American Institute of Electrical Engineers on the deceptively simple topic of 'A New System of Alternate Current Motors and Transformers'. The innocuous title hid a great deal. This was to be Tesla's first real opportunity to present his vision of the electrical future to an audience of fellow electrical engineers. It was his chance to show his mettle and establish his reputation as a serious player in the game of invention. Up to this point in his career, Tesla had been a relative unknown. Few even in his lecture audience would have recognized his name. The choice of platform for this public appearance had been entirely deliberate. The backers Tesla had acquired since leaving the Edison Machine Works had decided that the time had come to make him into a public figure. He had inventions and a vision to sell, and to do that successfully he needed first of all to sell himself as an inventor to inventors.

Tesla promised his audience 'a novel system of electric distribution, and transmission of power by means of alternate currents, affording peculiar advantages, particularly in the way of motors, which I am confident will at once establish the superior adaptability of these currents to the transmission of power and will show that many results heretofore unattainable can be reached by their use; results which are very much desired in the practical operation of such systems and which cannot be accomplished by means of continuous currents'.[1] It was an ambitious promise, but one that he was confident he could deliver. The current generated by electrical generators was alternating current; the current that actually worked electric motors was alternating current. His system was better than existing direct current systems because it was simpler and more elegant. It did away with the need for clumsy commutators to turn alternating into direct currents and back again.

The bulk of Tesla's lecture consisted of descriptions of engines that he had designed that operated directly with alternating current. He had brought some models along to show off as well. He showed his audience that it was indeed possible to make engines that ran directly from alternating currents – and that they were more efficient than conventional engines. He gave practical details about how to arrange magnets and armatures to best effect, and expounded on the theory behind his invention. This was a crucial aspect of Tesla's self-presentation. Not only was he an excellent and creative practical engineer – or that, at least, was how he wanted his audience to see him – but he was a sophisticated theoretician as well. That was what gave him an edge. He combined the practical know-how of a former Edison employee with the theoretical flair that resulted from a European university education in physics. It was one of the things that distinguished him from almost everyone else in that lecture theatre.

Tesla's performance was well received – and the reception had clearly been carefully orchestrated by Thomas Commerford Martin, the editor of *Electrical World* and a prominent member of the institute, who had been responsible for organizing it. He made sure that the discussion was opened by William Anthony, who had been professor of physics at Cornell University until the previous year, and had already seen Tesla's engines in action. Anthony's testimony was authoritative, and he could assure the audience that Tesla's claims about his engines' capacities were well founded. He thought that 'on first seeing the motors the action seemed to me an exceedingly remarkable one', he told the gathering. The 'little motor that you see here gave us about half a horse power, and had an efficiency of something above fifty per cent, which I considered a very fair efficiency for a motor of this size, as we cannot expect with such small motors to get as high efficiency as we can with larger ones'.[2]

The inventor Elihu Thomson took advantage of the opportunity to remind the audience that Tesla was not the only inventor of alternating current motors present. He had 'as probably you may be aware, worked in somewhat similar directions and towards the attainment of similar ends'.[3] Tesla in turn was quick to acknowledge Thomson's achievements. At the close of the meeting, after all the other papers had been delivered, the chairman interjected that he had 'another announcement to make in regard to Mr Tesla's motor. I believe that this motor – Mr Tesla can correct me if I am not right – is the first good alternating current motor that has been put before the public anywhere – is that not so, Mr Tesla? And he says that the system can be seen in practical operation at 89 Liberty Street, second floor, and he invites you all to come and see it.'[4]

So what had Tesla been doing between his abrupt parting of the ways with Edison and his companies in 1884 and his triumphant appearance at the American Institute of Electrical Engineers a few years later? His performance there was the culmination and a vindication of a long struggle to establish himself. Tesla had abandoned Edison with high hopes. He had been working on developing an arc lighting system for Edison – it was Edison's failure to take up the system that led to Tesla's departure – and now he wanted to exploit his work for his own benefit. He soon acquired financial backing to set up the Tesla Electric Light and Manufacturing Company. His backers, Robert Lane and Benjamin Vail, had been looking for ways of breaking in to the profitable business of electric lighting and Tesla's arc lighting system seemed to offer just the opportunity they had been looking for. Over the next few months Tesla developed a number of patent applications to protect his new system. By 1886 the Tesla arc light system was in use in the town of Rahway in New Jersey, where Lane and Vail were based.

Everything seemed to be going well for Tesla's future. On 14 August that year the New York trade journal the *Electrical Review* published a laudatory front-page article describing his arc lighting system. It described how 'Mr. Tesla, the inventor, has obtained broad patents on the regulation of a dynamo machine on entirely novel principles. This method of regulation secures advantages in the way of economy and safety that he is confident are peculiar to this system alone.'[5] But disaster soon followed. Lane and Vail abandoned him and the company, leaving him with little more than some worthless share certificates. He was reduced to working at odd engineering jobs and even working as a manual labourer, digging ditches. His salvation came thanks to another patent application he had filed in 1886, for a thermomagnetic motor. Intrigued by the prospect, two other businessmen, Alfred Brown and Charles Peck, offered him a partnership and rented a laboratory for him in New York where he could continue to work on his inventions.

Tesla was learning the hard way that there was more to remaking the future than mere invention. A successful life of invention needed financial backing, and Tesla needed to learn not only how to attract potential financiers but how to find money-men that he could trust. Tesla would later describe Brown and Peck as 'the finest and noblest characters I have ever met in my life'.[6] They not only provided him with resources and a laboratory but offered him some freedom of action too. But even they needed to be persuaded that Tesla's dreams of AC dynamos and motors could be made real in a way that met the needs of the growing electrical industry. Tesla might want to remake the electrical world from the ground up, but Brown and Peck needed to show a profit – and that meant making machines that could be adapted to already existing systems.

What Tesla really wanted to do was develop an entirely new system of alternating current generation, transmission and use. He had been thinking about how to do this – and in particular about how to design an entirely new kind of alternating current motor – since his student days. His fantasy was to make use of a number of alternating currents to create a rotating magnetic field. As the field rotated, it would make the motor's rotor turn as well. With Brown and Peck's support Tesla proceeded to develop his ideas for a polyphase motor along those lines. They also hired a patent lawyer to work with Tesla in preparing his patent application. The lawyer, Parker W. Page, had a particular interest in electric motors, being the son of Charles Grafton Page, who had been at the forefront of attempts to build practical electric motors for locomotion during the 1840s.[7]

What Tesla wanted to patent was his entire system, not just the polyphase motor. This really was a matter of rebuilding the electrical world. If Tesla had his way then an entirely new network of AC transmission would be needed – and it would be expensive, needing up to six wires connecting generators to motors. Page, on the other hand, realized that Tesla needed to curb his ambitions, for the moment at least. He needed to develop a motor that could be adapted to existing networks of AC transmission. Copper was expensive, and potential customers were unlikely to want to bear the cost of laying additional cables. Tesla had to be persuaded that as well as developing and patenting his comprehensive system he had to develop a polyphase motor that could work in the electrical world that already existed. That was the motor that Tesla eventually put through its paces in front of the American Institute of Electrical Engineers in May 1888.

That performance was carefully staged. Tesla had already shown his motor to Professor Anthony and he was primed to ask the first question. With Page, Brown and Peck's endorsement he had acquired some other powerful advocates as well, such as Thomas Commerford Martin. Tesla badly needed the support of men like Martin if he had aspirations to be anything more than just another inventor. He needed that support particularly since it must have seemed to many at that meeting of the American Institute of Electrical Engineers that Tesla, however ingenious his motor was, was backing the wrong electrical horse. The canny money was on direct current systems, not alternating current ones. If Tesla was to be successful, he really did need to persuade the industry to change direction in quite a fundamental way. As far as most electrical engineers and investors were concerned the shape of the future was already determined, and it was a future powered by direct, not alternating current. Direct current was, superficially at least, the simplest system. Electricity flowed uniformly in one direction. In alternating current systems, the electricity reversed direction at regular intervals, typically several times a second. Both systems had their advantages and disadvantages.

Edison, of course, was one of the leading proponents of direct current systems. Edison had been thinking about and developing ways of distributing electricity at the same time he and his employees at Menlo Park had been developing the incandescent light bulb. This was one feature of the business of inventing the future, at least, where Tesla and Edison agreed. It was not enough to invent generators, motors and other ways of consuming electric power in isolation. Everything had to fit together. That was the problem that Tesla had to find ways to overcome. By the end of the 1880s much of the work of developing and consolidating electrical networks of power was well under way – and it was being done to Edison's design rather than Tesla's. Edison had decided relatively early on in this process that direct current systems were the best way to proceed. The world of direct current was already expanding rapidly just as Tesla was trying to paint his picture of an alternative electrical future.[8]

For Edison and his backers – just as it would be for Tesla and his backers – the most important feature of this electrical future was that it should be profitable. When Edison set up the Edison Electric Light Company at the end of 1878 his fellow board members included some very prominent financiers. Their financial backing was essential to Edison's success. They provided him with the resources he needed to experiment and find the right combination of elements for his system. But they wanted a return for their money. Electric light already had a competitor in the form of gas lighting systems. Edison had to convince his investors that he could see off this competition from a well-established and profitable rival system. Costs mattered, not only for developing the apparatus, but for manufacturing, installing and maintaining the new network of generators and wires. From the beginning, the need to generate a profit was central to the way that Edison imagined his system.

For this and many other reasons it was clear to Edison that the future lay in direct current systems. Their characteristics were better understood, and they seemed more versatile. Edison understood that to make a success of his ideas the new electrical networks would need to do more than simply provide the power for his incandescent lights, for example. Light was only needed at night – so what would his dynamos power during the day? To be economically viable they would need to be used to power machinery of various kinds. While existing generators (such as the Gramme dynamo) necessarily produced alternating current, this was invariably converted by commutators into direct current for transmission and use. Since the earliest attempts to develop commercially economic electric motors during the 1840s such devices had worked by direct current. So, by the 1880s, direct current already had a long history. It was tried and tested – and therefore relatively risk-free – technology.

It is worth bearing in mind that the direct current system of electrical transmission and use that Edison envisaged actually owed a great deal to the networks of gas lighting systems with which it was to be in competition. Just as the gas was generated in a central facility, so would electricity be. Just as gas was distributed throughout towns and cities through a network of pipes, so would electricity be – in fact early plans for electricity distribution assumed that the wires would run underground like gas pipes rather than through overhead cables. Like gas, this was also to be a system designed to distribute electricity to relatively large towns and cities. Direct current electricity could not be transmitted over very long distances before becoming uneconomical, with too much power being lost in transmission due to the resistance of the wires. This kind of electrical power could only ever really be an urban phenomenon.

By the end of the 1880s urban electricity really was starting to become common. Edison had opened his first electrical power station on Pearl Street in New York in 1882. It was in the heart of the city's financial district – and deliberately so, since Edison and his backers wanted to be sure that potential investors could see what was happening on their doorstep. On 4 September Edison himself turned the lights on in the financier J.P. Morgan's offices and the press pronounced the wizard of Menlo Park's latest enterprise as 'eminently satisfactory'.[9] When the plant opened it had

52 customers, but the number soon multiplied to several hundred. Underground cables connected the central generating station where Jumbo the dynamo (named after P.T. Barnum's famous elephant) churned out its power, to domestic houses and commercial businesses. Within months Pearl Street was powering a network of several thousand incandescent lights. Edison's success persuaded others both elsewhere in the United States and in Europe that his system of direct current electricity generated centrally and distributed at a relatively low voltage was the shape of energy to come.

Edison's system spread in Europe both by careful promotion and forging business relationships with potential backers across the continent. Edison had mounted a spectacular exhibition of his incandescent lights at the Paris exhibition a year earlier. William Henry Preece, the British Post Office's chief engineer, had marvelled at how the system 'had been worked out in detail, with a thoroughness and mastery of the subject that can exact nothing but eulogy from his bitterest opponents'.[10] Even before the Pearl Street power station was opened, Edison's London agents had established a power station at Holborn Viaduct on exactly the same system. The German engineer Emil Rathenau was another Paris convert. He bought the German rights to Edison's patents, scenting an opportunity to snatch the technological lead in the German electrical industry from Werner von Siemens and his Siemens & Halske company. Siemens & Halske were soon manufacturing incandescent lights under licence from Edison's German company.

Not all Europeans embraced the direct current system nevertheless. In London, when the Italian engineer Sebastian de Ferranti started developing his plans for what would become the Deptford Power Station, his vision was of a power station generating alternating current at an entirely unprecedented scale. Work on the project started in 1887 and was completed two years later. Even while work was in progress it was clear that Ferranti was attempting an entirely novel approach to generating power. The Deptford Power Station was not designed to provide electricity for a district or a town. It was designed to produce enough power for the whole of London. Rather than transmitting electricity at a potential of a few hundred volts, electricity from Deptford would be transmitted at 10,000 volts. This was an entirely different scale of working that embodied a wholly different vision of what the future electrical world would look like. The electrical press marvelled that 'the new electrical machinery is so enormous compared with anything in existence, that it may be deemed an entirely novel creation, a monument in the confidence reposed by the directors in their engineer, Mr. S. Z. de Ferranti.'[11]

Ferranti's ambition did not go unnoticed on the other side of the Atlantic. The *Electrical World* called the Deptford Power Station a 'herculean enterprise'.[12] It was certainly an enterprise towards which Tesla would have been deeply sympathetic – and interested as he was in developing his own alternating current system at exactly this time, he must have been aware of what Ferranti was doing. Ferranti's efforts must also have attracted the attention of George Westinghouse and his Electrical Manufacturing Company. Westinghouse was a relative newcomer to the electrical world and was investing heavily in the possibilities of alternating current systems and devices. He was certainly interested in Tesla's polyphase motors and the patents he had acquired for alternating current devices. A few months after Tesla's demonstration of his motors at the American Institute of Electrical Engineers, Westinghouse bought Tesla's patents. Tesla himself started working for Westinghouse to turn his motors from demonstration devices into real, practical, working machines. By the end of the 1880s the stage was set for what would be an increasingly vicious commercial and technological battle of the systems.

Chapter 7

Building Tomorrow

The 1880s was the decade when electricity and electrical power properly came of age. Up until now, for most people electricity had been a spectacle to be wondered at in exhibitions and popular lectures. They might have been bedazzled by electric light at the theatre. Telegraphy might play an increasingly important role in nineteenth-century lives, but beyond encounters with telegraph clerks while dictating messages, most of the public had little experience of the electrical machinery behind the scenes. Maybe the more adventurous had experimented with electrotherapy. Electricity was the stuff of the future, and outside their imaginations most people needed to visit the places where the future was put on show to see electricity in action. By the 1880s, however, electricity was starting to become the stuff of today as well. In urban Europe and North America, it was increasingly leaking out of laboratories, exhibitions and lecture halls and finding a place on the streets and in people's houses.

The famous cartoon of Edison dressed as the Wizard of Menlo Park captures the ways in which the paraphernalia of electricity was starting to become familiar. The *Graphic*'s editor presumably expected the newspaper's readers to recognize what they saw – even the mirror galvanometer on Edison's hat. The instrument had originally been developed by the Scottish natural philosopher William Thomson for use on the Atlantic telegraph, and Edison had invented his own version of the device in connection with his telegraphic work.[1] Across the Atlantic, an equally iconic image of the Italian electrical engineer Sebastian de Ferranti suggests the increasing visibility of the electrical future. Ferranti was pictured as the Colossus of Rhodes – 'The Modern Colossus' – straddling the Thames just as the original statue of the sun god Helios had straddled the entrance to the harbour at Rhodes, but holding aloft a dynamo in one hand and an incandescent light in the other.[2] It would not be long before Tesla was the subject of this kind of portrayal as well.

A cartoon in *Punch* offers a tantalizing glimpse of the electrical future at the beginning of the 1880s. In it, Mr Punch dreams of the 'coming force' and a new electrical world.[3] Centre stage is

Sebastian de Ferranti as 'The Modern Colossus' straddling the Thames. (*Electrical Plant*, May 1889)

a giant dynamo on a cart being pulled along by horses adorned with incandescent electric lights. In the cavalcade behind, along with electric sprites and fairies are 'frozen beef from Australia revived by electricity' and a man relaxing on his electric bicycle and reading the 'Electric Times'. There are boxes of 'fruit, flowers and vegetables, grown by electricity' and advertisements for 'Christmas turkeys, hatched by electric heat'. Cowering and fleeing from the electric procession are a motley crew of chimney sweeps and gas lighters. The days of coal and gas, of steam engines and gaslight, were clearly numbered. Mr Punch, slumbering in the corner, has dropped a paper advertising the Electric Exhibition at the Crystal Palace – the source, presumably, of his electric reverie.

This was satire, of course. But it was satire with its feet on the ground. Much of what the cartoon showed (apart from electrical fairies and electrically resurrected cattle, at least) already existed. *Punch*'s readers might have already seen them at the Crystal Palace. Following the successful conclusion of the Great Exhibition of 1851, the spectacular building of glass and steel that housed it was taken

Mr Punch dreaming about 'The Coming Force' during the Electrical Exhibition at the Crystal Palace. (*Punch Almanack*, 1882)

down and rebuilt at Sydenham. There, the Crystal Palace became London's premier exhibition site – and just the place to show off the electrical future. The Crystal Palace's Electric Exhibition was itself an indication of the electrical shape of things to come. The exhibition was primarily devoted to exhibiting dynamos and electric light (Edison's Electric Light Company was there, for example) – and it was explicitly aimed at demonstrating that these were now commercial technologies. The *Daily News* enthused shortly after the show opened that the exhibition was 'becoming daily more interesting and more popular'. The 'crowds that throng the stalls' were proof of 'how eager the public is to learn'.[4]

Organized by the Crystal Palace's managers in an attempt to capitalize on the success of the electrical exhibition held in Paris the previous year, this display of the latest electrical technology was meant to demonstrate just how quickly the world was moving towards its electrical future. There was widespread agreement that the exhibition showed that progress in developing electric light and power had been substantial since the Parisian show. As the exhibition approached its end, even Charles Siemens agreed that the exhibition 'marked very considerable progress since its predecessor in Paris' and that 'they had a much larger show and a more complete demonstration of the efficiency and usefulness of the electric light.'[5] Electrical lights were not the only exhibits, of course. The War Office mounted a display of electric mines. For those whose interests were more domestic, there was an exhibition of 'a delightful little dynamo motor for working an ordinary sewing machine'.[6]

Surprisingly perhaps, there was no display of electrical locomotion at the exhibition. Londoners would not have to go far to see such an exhibition, however. On 4 March that year, just as the Crystal Palace exhibition was getting under way, the electrical engineer Radcliffe Ward demonstrated the first public electric tramcar at the North Metropolitan Tramways Company in Leytonstone.[7] There had been demonstrations of model electric locomotives before – Tesla may have known about Siemens & Halske's exhibition of a model electric railway at the Berlin Exhibition in 1881 – but Ward's demonstration meant business. It launched the Faure Accumulator Company, established to manufacture the electric accumulators (rechargeable lead-acid batteries) that would power the tramcars. Electric bicycles like the one lampooned by *Punch* had been on show at the Paris International Exhibition in 1867, and an electric tricycle toured the streets of Paris during the Electrical Exhibition there in 1881. By 1884, the English electrical inventor Thomas Parker was driving around in an electrically powered car.[8]

None of this can have escaped the attention of Tesla, dreaming as he was of the possibilities of electricity. By the time Tesla arrived in America in 1884 the electric tramcar had arrived as well. The first commercial electric-powered trams were operating in Cleveland Ohio by 1884, operated by the East Cleveland Street Railway Company. The same year, the Sprague Electric Railway & Motor Company was established by Frank J. Sprague, another of Edison's former Menlo Park employees. Where Ward's trams were powered by arrays of batteries, Sprague's locomotives gained their power from poles connected to overhead wires. Electricity was starting to change the urban landscape. By the end of the 1880s, city streets in Europe and North America were festooned with wires. Overhead cables carrying electricity for domestic lighting and cables to power electric trams joined the webs of telegraph and telephone wires that already criss-crossed city skies.

As we have seen, as early as 1877, the *New York Sun* announced the imminent invention of another new electrical communication technology – the telectroscope – that seemed to take the telephone a giant step further by making it possible to see at a distance. A similar instrument was announced by *The Times* of London a couple of years later. 'M. Senlecq, of Ardres, has recently submitted to the examination of MM. du Moncel and Hallez d'Arros, a plan of an apparatus intended to reproduce telegraphically at a distance the image obtained in the camera obscura,' it reported.[9] The report was widely circulated. A year later, the *Electrician* announced that the telectroscope had 'everywhere occupied the attention of prominent electricians who have striven to improve on it.'[10] Seemingly, wires carrying sight as well as sound were soon to be added to tangle of cables adorning city streets.

There never was an actual telectroscope, though rumours that such a device had just been invented persisted for the rest of the century, but the persistence of the rumours – which surfaced repeatedly throughout the remainder of the century – shows the extent to which the possibilities of electricity were increasingly tangible. More explicitly fictional, the electrical technologies that featured in Edward Bulwer-Lytton's *The Coming Race* gained traction with his readers because of their familiarity. The underground world he pictured seemed all the more plausible because the machines the vril-ya used to dominate their environment were already part of the real world's electrical future. In Europe and North America enthusiasts speculated about the possibilities of electrical agriculture, electrical control of the weather, electrical transport – and electrical weaponry. This was a world of fantastic possibilities in which the revolutionary potential of electricity seemed endless, and the line between fact and fiction hard to draw.

Or, as the cynics would have it, 'electric savants, unlike most men of science, are doing their thinking aloud, performing experiments in public, talking to each other across continents and in the ears of half mankind, showing instruments which they confess are imperfect, exhibiting processes which are acknowledged to be merely tentative, securing patents which are defended as only "precautionary," and in many instances letting drop hints as to the methods by which they are inquiring, and the results they barely hope to obtain, which on other subjects would arouse in their hearers a sense of angry tedium.'[11] There were clearly times when the electrical dream seemed too good to be true. This electrical dreaming suited writers of scientific romance nevertheless. Peppering their pages with electricity was the best way of making the fantastic seem real to their readers.

So as early as the 1860s Jules Verne's heroes descended to the centre of the Earth with the help of light generated by their 'ingenious Ruhmkorff lamps'. A few years later, Captain Nemo was running the *Nautilus* with the help of 'a powerful agent, responsive, quick, and easy to use, pliable enough to meet all our needs on board. It does everything. It supplies light and heat for the ship and is the very soul of our mechanical equipment. That agent is electricity.' Verne even detailed how Nemo produced electricity from the seawater itself.[12] The future as written by Albert Robida in 1882, just around the time that Tesla arrived in Paris to work for Edison's company, was even more electrical. In Robida's Parisian twentieth century there was very little that was not electrical. The inhabitants of twentieth-century Paris could even enjoy opera through their telectroscopes, although Robida called the devices telephonoscopes.[13]

Robida's future Parisians (the narrative was set 70 years later in 1952) lived, worked and played with electricity. They travelled underground on an electrically driven tube train or through the air in an omnibus-airship, guided from a landing tower crowned by an electric lighthouse. They talked to each other through telephonographs. Their entertainments were produced by electricity. Even their dinners were electrically delivered to their apartments – with hilarious consequences. Their daily working lives were regulated by electric clocks and their banks protected by electric alarms. The city was defended by barrages of electric weaponry. Like the *Punch* cartoon, this was satire firmly grounded in reality. The future that Robida poked fun at (and his target was presumably the previous year's Parisian exhibition) was a future that he and his readers fully expected to take place, and much of the humour in the novel consisted of exploring the consequences of extrapolating post-imperial Paris's social mores and values into that imagined world of electricity everywhere.

In the United States, the Steam Man of the Prairies, who had been a figure in Boy's Own adventure stories from at least 1868, was by the early 1880s in the process of being replaced by the even more impressive Electric Man. This was electricity (and electrical invention) as a vehicle for manifest destiny. The Steam Man and his electrical successor featured in story after story in which they conquered the plains, guided by their intrepid young inventor Frank Reade Jr, and having conquered America they set out to conquer the world. The Electric Man was 'a Samson in physical strength' running on a 'powerful electric battery' and with eyes that produced 'a light that equals the noonday sun'.[14] He was soon joined by an Electric Horse, fully equipped with guns and electric torpedoes. Written by the hugely prolific Luis Philip Senarens, the Frank Reade stories captured nicely what electricity was coming to mean for America, at just around the time that Tesla was arriving in New York to make his fortune with invention. The inventors that appeared in these stories were indomitable individualists, and in due course, his name made, Tesla would feature in at least one such tale of inventive adventure himself.

The reality of electricity in the 1880s was rather different. The business of making sure that the tentacles of electrical progress spread in the right directions was increasingly a matter for national governments and for large corporations. Electricity could not be allowed to simply flow unchecked everywhere. It needed regulation. That had been one of the important lessons of the Atlantic cable fiasco a couple of decades earlier. The first cables had failed in part because there were no commonly agreed standards between manufacturers and between different countries. Cables made by one manufacturer in London might have quite different electrical properties to cables manufactured in Birmingham, let alone ones made in Berlin, Paris, or New York. If the world was going to be made electrical that kind of anarchy could not be allowed to prevail.

The International Electrical Congress that took place at the Paris Electrical Exhibition in 1881 was one sign of the new world order. At the meeting, eminent men of science from across Europe gathered to define the electrical world. Two giants of electricity, William Thomson and Hermann von Helmholtz – both vice-presidents of the congress – went head to head over the question of how to define electrical units. It was a major triumph for British electricity, and the growing British electrical industry, when the meeting agreed to accept the definition of the ohm (the unit of electrical resistance) proposed by the British Association for the Advancement of Science, based on experiments conducted at Cambridge's Cavendish Laboratory by James Clerk Maxwell and his successor as director of the laboratory Lord Rayleigh. Agreements like this were a vital element in getting the electrical future going. Imposing standards and defining terms were essential for making sure that engines manufactured in one place could be trusted to work reliably anywhere else those standards were accepted.[15]

By the end of the 1880s electricity really was embracing the globe. Following the eventual successful completion of the Atlantic cable in 1866, undersea cables were soon proliferating, connecting Europe with overseas colonies. By the mid-1870s telegraph cables connected Europe to India, south-east Asia and China, and spread onwards to Australia and New Zealand. By the end of the decade a cable was laid down the east coast of Africa, followed by another down the west coast by 1885.[16] While these cables were laid by private companies, European governments took a keen interest in the proceedings. Swift communication with the peripheries of empire made imposing the imperial will easier to manage. In the United States the telegraph, like the railways, was a vital tool for westward expansion. Writers of scientific romance might fantasize about how electrical weaponry would transform future warfare. By the 1880s telegraphy and telephony were already changing the face of battle.

So by the time Tesla set foot on American soil in 1884 the electric future already seemed to be taking shape, and the electrical inventors who seemed to be shaping that future were starting to become household names. Edison, of course, led the field in such self-promotion, but he certainly was not the only electrical entrepreneur to realize that publicity – and more than a touch of showmanship – were essential elements of the game of invention. Inventors like Elihu Thomson knew that exhibition was a key part of the process of getting their inventions out of the workshop and into homes and workplaces. But just like Edison, men like Thomson also understood that invention was a business. Thomson in 1883 gained financial backing to establish the Thomson-Houston Electric Company to manufacture and market inventions such as his arc lighting system and alternating current dynamo.[17]

Inventing the future was by the end of the 1880s turning into a cut-throat, competitive affair. Ambitious inventors jostled for opportunities to set their wares before the public. As Tesla himself was in the process of discovering, inventive genius was not enough to guarantee success. Rival inventors and corporations were engaged in fierce competition to get their products on the market and to attract investors. By the end of the decade one battle in particular was getting hotter – and it was a battle in which Tesla himself would be a major combatant. This was the war of the currents being bitterly fought between Edison's Electric Light Company and the Westinghouse Electric Company over the respective merits of their rival direct current and alternating current systems of electric power transmission. Tesla, with his vision of an alternating current world, and with his alternating current motors to sell, would be in this war on Westinghouse's side.

PART 3

SCIENTIFIC SHOWMAN

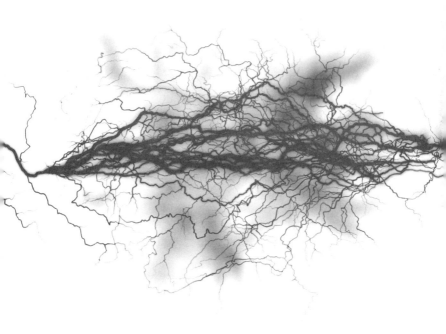

CHAPTER 8

The Business
of Invention

There does not appear to be an account of Tesla's first encounter with George Westinghouse, but it must have been a memorable meeting. Tesla was certainly impressed by the man. Recalling Westinghouse after his death, he remembered the 'tremendous potential energy of the man ... A powerful frame, well proportioned, with every joint in working order, an eye as clear as a crystal, a quick and springy step – he presented a rare example of health and strength.'[1] Westinghouse, when Tesla met him, was in his early forties – just a decade or so older than Tesla himself. He already had a solid track record in the business of invention, though his interest in electricity was relatively recent. After dabbling in direct current (DC) power distribution for a couple of years, he had grasped the possibilities of alternating current distribution, and particularly the potential economies of scale that could be achieved. Over the next few years, he invested heavily in developing his own alternating current systems.

Westinghouse wanted the patents that Tesla had taken out for his polyphase motors for a number of reasons. The Westinghouse Electrical Manufacturing Company needed a viable alternating current motor to offer to their customers. His own chief electrician, Oliver Shallenberger had been working on such a prospect, and Westinghouse had already bought the rights to the alternating current motor developed by the Italian electrician Galileo Ferraris. Tesla's patents therefore represented not only an opportunity but, viewed from another angle, a potential threat to Westinghouse's ambitions to dominate the market in alternating current systems. Buying the patents was the best way of neutralizing the threat as well as realizing the opportunity. Convinced by Shallenberger that Tesla's motor was already far superior to his – 'the best thing of the kind I have seen' – Westinghouse not only agreed to buy the patent, but he agreed to hire Tesla as well.[2] Tesla would move to Pittsburgh to work with Westinghouse's engineers on improving the motor and on adapting his invention to fit the practical needs of the electrical marketplace.

For the second time since he arrived in America, Tesla found himself working for another inventor. His situation now, however, was quite different from the lowly position he had occupied in Edison's business empire. Tesla was one of Westinghouse's most important assets. Arriving in Pittsburgh in July 1888, Tesla spent the next several months working at improving his motor, filing fifteen patent applications in the process. Much of this work involved experimenting to find out what the best materials for producing the motors on a commercial basis would be, and adapting the design to make them easier to manufacture. There were disagreements from the beginning, nevertheless. Tesla thought his motors worked best at relatively low frequencies, but Westinghouse and his engineers wanted motors that operated at the higher frequencies their systems already used. Tesla was persuaded to adapt his design – a lesson in the practicalities of bringing new inventions to the market.

Westinghouse initially hoped that the Tesla motors would be sold to be used in streetcars working from overhead cables. There was also a market for the motors to be used in the operation of mining machinery and many were shipped out for that purpose. But barely a year after starting to work for Westinghouse, Tesla was thinking about leaving. He was unhappy with the way in which commercial considerations were being given priority over technical ones in bringing his motors to the market. Despite his leading position in the Westinghouse business, he was learning the hard way yet again that there was more to the business of invention that invention itself. In August 1889 he resigned his position and set off for Europe – his first return since arriving in America in 1884. Tesla was discovering that the discipline of bringing his inventions to the marketplace was a difficult one for him to master. It was time to take stock.

August 1889 was in any case a good time for an electrical inventor to return to the Old World – and to Paris in particular. Tesla was travelling as part of a delegation by the American Institute of Electrical Engineers to the International Congress of Electricians that was being held in conjunction with the Universal Exposition being held that year in Paris. His membership of the delegation was itself testimony to the way his star was now rising in the firmament of American invention. It was at this congress that the watt and the joule were adopted as the internationally recognized practical units for work and power respectively. Comparing the Paris exhibition with the Viennese one just over a decade previously, the United States commissioners in their report noted that since then 'the telephone, the microphone, the arc light, the incandescent light, the practical systems of distribution of light, the accumulator, and other fundamental inventions forming the present industry, began to come into use.' Paris therefore represented 'the extent of the present electrical industries that have made such wonderful progress in so short a time, not much over ten years'.[3]

There was certainly a great deal going on at the Paris exhibition to fascinate and intrigue someone like Tesla. It offered, as Carl Hering, the US commissioner for electricity had noted in his report, a graphic illustration of just how rapidly the world was moving towards an electrical future. A few months before Tesla arrived, the Eiffel Tower had been opened to the public. It was a tangible symbol of progress. During his Parisian stay, Tesla had plenty of time to explore the Universal Exposition's electrical exhibits. Edison was there as well (though there is no record of any encounter) and working hard at promoting his inventions. It could not have escaped Tesla's attention just how ubiquitous Edison's productions were throughout the exhibition. His dynamo was one of the largest ones on show. Westinghouse, by contrast, had nothing to exhibit. Hering, the US commissioner, noted in his report that compared to the continuous current systems on display, the 'exhibits of alternating-current dynamos were very few, and of these there were only a very few of interest'.[4] It was an object lesson in the importance of showmanship.

It must have looked to most observers at the end of the 1880s as if the electric future would indeed operate on continuous currents. Edison and his companies dominated the Paris exhibition – they certainly dominated the American exhibits there. There might still be a role for alternating current systems for specific purposes such as mining, but the urban world of the future – and the American urban world in particular – would be a continuous current one. Nevertheless, it is clear that Edison thought that alternating current represented enough of a threat to his vision of the future to require his active opposition. Throughout the 1880s there had been periodic discussions regarding the possibility of using electricity as a safer, more certain, and more civilized means of execution. Edison, originally an opponent of the death penalty, scented an opportunity. Electricity, he declared, really would be a safer way of killing than the rope, he declared – as long as the electricity used was alternating current.[5] The basis for Edison's claim was that alternating current was usually transmitted at far higher voltages than direct current, so that contact with the wires was far more likely to be lethal.

Appearances aside, it is clear that alternating current systems – with Westinghouse as one of their chief advocates – were posing an increasingly serious threat to Edison's businesses by the second half of the 1880s. Edison, with his flair for showmanship, was convinced that the best way of countering the threat was through spectacle. As more and more overhead cables criss-crossed the sky above the streets of American cities, there were certainly fears, fuelled by lurid press reports, about electrical safety. Edison anticipated that a campaign playing on those fears would destroy the public's and investors' appetite for alternating currents. Soon, Edison's protégé Harold P. Brown was conducting widely reported experiments on the dangers of alternating currents, killing animals on stage with electricity. In his *Comparative Danger to Life of the Alternating and Continuous Electrical Currents* (1889), Brown was quite clear that alternating current was the current that kills.[6] If a new and more scientific means of execution were required, alternating current was the answer. Edison even quipped that the term 'to westinghouse' should be employed to describe the process of electrical execution.

Exhibiting the dangers of the high voltages needed to make the long-distance transmission of alternating current economic was central to Edison's campaign against Westinghouse. As Westinghouse attempted to establish his alternating current system in New York, directly encroaching on Edison's territory, Edison lobbied state legislators to limit the maximum voltage allowed in overhead transmission lines. He invited journalists to his laboratory to see for themselves the deadly consequences of contact with high voltages. Edison's lobbying and Brown's very public demonstrations of the dangers of high-voltage alternating current (to stray cats and dogs, at least) had the desired effect. A campaign to adopt electricity as a means of execution in New York had been under way for some time. Campaigners argued that death inflicted in such a way was entirely painless and humane compared to the uncertainties of hanging. On 1 January 1889 a bill introducing electricity as the sole means of execution in the state of New York was passed into law. Further intense lobbying ensured that the precise method of execution would employ alternating current.

The first electrical execution took place in Auburn prison in New York on 6 August 1890. When the bill had been passed more than eighteen months earlier little consideration had been given to the actual method of achieving death by electrical means and it took some time for experimenters to settle on a chair to which the victim could be secured as the best way of proceeding. The Westinghouse generator installed at the prison also needed to be tested to make sure that it generated a killing current. The electric chair's first victim was William Kemmler, convicted of the murder of his common-law wife Tillie Ziegler. Once it became clear that Kemmler would be executed by electricity rather than hanging, attorney William Bourke Cockran embarked on a desperate series of appeals (financed, it was rumoured, by George Westinghouse, equally desperate to avoid having his alternating current system damned by association with the deed) to overturn the sentence of death by electricity as cruel and unusual punishment.

The execution itself was a fiasco. Far from being a humane and painless method of killing, electrocution turned out to be a protracted and grisly process. In the end, it took two attempts to finish off Kemmler. The *New York Times* reported that 'the witnesses were so horrified by the ghastly sight they could not take their eyes off it … Blood began to appear on the face of the wretch in the chair. It stood on his face like sweat … An awful odor [*sic*] began to permeate the death chamber, and then, as though to cap the climax of this fearful sight, it was seen that the hair under and around the electrode at the base of the spine was singeing. The stench was unbearable.'[7] The newspaper was in no doubt that the first electrical execution should be the last: 'this single experiment warrants the prompt repeal of the law.' For Westinghouse, of course, the damage had already been done. Edison's campaign to make alternating current the current that kills had been successful. Increasing numbers of professional electrical engineers might recognize that alternating current systems were technically superior to their direct current rivals, and more economic, but they were getting a hammering in the press.

By 1890 Westinghouse and his alternating current systems were in deep financial trouble. The rapid expansion of the alternating current system that Edison had found so threatening had been financed by heavy borrowing. The alternating current business had survived Edison's attempts to undermine it and was booming, but the boom meant that more investment was needed to finance the expansion. Westinghouse himself lent his own company $1.2 million in an attempt to keep it solvent. Disaster struck before the end of the year when Barings Brothers in London collapsed, triggering a financial panic that induced many of the Westinghouse Electric Company's creditors to demand early repayment of their loans. The company was forced into receivership. Even Tesla felt the blow, despite the fact that he was no longer a company employee. His contract with Westinghouse, giving him a share in the royalties from his patents, was cancelled as new investors insisted that the company reined in what they regarded as reckless investment in untried inventions. Tesla had little choice other than to agree to ending the contract, since if Westinghouse lost control of his own company, he was unlikely to see any further profit from it in any case.

But Westinghouse was not the only one in trouble. Edison's control of his own company was starting to look increasingly insecure as well. Not all his investors were as convinced as he seemed to be that the future of electricity lay with continuous currents. His apparently obstinate refusal to consider the possibility of adopting alternating currents, particularly in view of developments on the other side of the Atlantic such as Ferranti's achievements at the Deptford Power Station, was raising eyebrows. The adoption of alternating currents across Europe offered tangible evidence that the alternative system really did work and that concerns about its safety were overblown. Properly dealt with, alternating current was not the current that killed after all, despite the lurid press reports. In the end, in February 1892, the Edison Electric Company merged with the Thomson-Houston Electric Company (Elihu Thomson's business and already active in alternating currents) to form the General Electric Company. Edison was soon an increasingly marginal figure in the new company's future.

Tesla, in the meantime, clearly took full advantage of his trip to Europe and his visit to the Paris Universal Exposition to do some hard thinking about his own future. What kind of inventor did he want to be – and what kind of contribution did he want to make in future to the business of invention? His experiences with Edison and with Westinghouse seemed to indicate that he was not well suited to working for others. Committed as he was to a life of invention, he needed to find ways to maintain his independence, and Paris offered Tesla exactly the opportunity he needed to look for a new direction for his work. There was certainly plenty on show in Paris to excite his imagination. It was during this trip to Europe that he heard at first hand about Heinrich Hertz's discovery of electromagnetic waves in 1888, providing experimental confirmation of James Clerk Maxwell's theories of electromagnetism.

Hertz's discovery seemed to open up a whole new field not just of discovery but of invention. Among Maxwell's disciples in Britain, the discovery that electromagnetic waves other than light could be generated and detected travelling through the ether was celebrated as conclusively demonstrating the existence of the electromagnetic ether that Maxwell had postulated. Maxwell had been adamant that it was impossible to conceive of energy travelling through space without there being a medium through which that energy could travel: 'whenever energy is transmitted from one body to another in time, there must be a medium or substance in which the energy exists after it leaves one body and before it reaches the other.'[8] Hertz's discovery meant that Maxwell was right, and like many others Tesla recognized that all kinds of prospects for invention might lie waiting to be discovered in the vortices of the electromagnetic ether.

Tesla, in fact, thought that he understood better than Hertz himself just what was going on in the crucial experiment. Hertz, like everyone else, understood that the key phenomenon in the experiment was the transfer of energy through the ether in the form of an electromagnetic wave. Tesla thought differently. Yes, the experiment demonstrated the existence of electromagnetic waves as Maxwell had predicted, but they were just a side-effect. What was really going on – and what generated most of the phenomena associated with the experiment – was a huge build-up of electrostatic charge in the electric field around the induction coil used to generate the Hertzian waves, as they were often called. Tesla did not think that much could be done with the waves themselves. That huge electrostatic charge building up in the ether around the induction coil was a different matter, however, and would become the main focus of his future researches.

Following his return to the United States, Tesla had plenty of possibilities to investigate as he contemplated future inventions. Investigating Hertz's discovery was one of the first things he did following his return from Pittsburgh to New York and moving, self-funded, to a new laboratory on Grand Street and assembling a small group of mechanics and electricians to help him. 'Dr. Heinrich Hertz's results caused a thrill as had scarcely ever been experienced before,' he recalled, and he 'fairly burned with desire to behold the miracle with his own eyes'.[9] Not satisfied with simply repeating what Hertz had done, Tesla proceeded to improve the apparatus. Instead of using a mechanical device to interrupt the current in his induction coil to generate a spark, Tesla used the high-frequency alternator that he had been developing. This meant that he could work the apparatus at far higher frequencies than Hertz himself had done.

It was while tinkering with Hertz's apparatus like this that Tesla developed what he called his oscillating transformer. There were three directions in which Tesla thought he might look for new inventions. There were 'the excessive electrical pressures of millions of volts, which opened up wonderful possibilities if producible in practical ways; there were currents of many hundreds of thousands of amperes, which appealed to the imagination by their astonishing effects, and, most interesting and inviting of all, there were the powerful electrical vibrations with their mysterious actions at a distance'.[10] With his new invention he was able to investigate the effects of high potentials and high frequencies at the same time. The oscillating transformer, or Tesla coil, was going to be the key instrument around which most of Tesla's activities would revolve for the rest of the century.

In the two years between his lecture to the American Institute of Electrical Engineers in 1888 and his invention of the Tesla coil in 1890 Tesla had transformed himself as an inventor. In 1888 he had been just another impecunious patentee. By 1890 he was a name to reckon with. In that process he had played a key role in transforming the electrical landscape of America too. Against the odds, the day of direct currents was over. In the end, it was a simple matter of economics. Direct current systems were only profitable in towns and cities where there were large enough populations to support them, since direct currents could only be transmitted over relatively short distances. Only alternating current systems could supply small town American with an electrical future and make money at the same time.

During those two years Tesla had learned the business of invention and discovered the role that best suited him. Working with Westinghouse had convinced him that he needed to be his own master. Tesla realized that he simply was not cut out for the compromises necessary in long-term collaboration. To remake the future as he wanted, he needed to be able to control his own destiny. What he needed from others was the financial backing to develop his ideas without interference. Tesla would work with Westinghouse again, but it would be on his own terms. With the invention of the Tesla coil, and his decision to devote himself to exploring 'the practically unknown regions' that the device promised to open up for him, he was set to transform the future and make himself a fortune in the process.

CHAPTER 9

Electrical Landscapes

B y the beginning of the 1890s the world really was beginning to look electrical. In European and American cities, the paraphernalia of electrification was everywhere. Electrical power lines criss-crossed the sky. Electric street cars, powered either by electricity from some of those cables or from phalanxes of batteries, were in the process of replacing horse-drawn omnibuses in many places. Electric street lights were a familiar sight, not just in large cities but in smaller towns as well. As early as 1880, the *Scientific American* recorded that the town of Wabash in Indiana was laying claim to being the first place in America entirely illuminated by electricity.[1] Town and cities embraced civic electric lights as symbols of modernity. Electric lighting was increasingly common in the houses of the well-to-do middle classes by the beginning of the 1890s. Domestic light fittings were ornate and elaborate at first, often designed to imitate the gas lighting they were replacing.[2]

The electrical engineer and Edison protégé Arthur Kennelly suggested in 1890 that 'the adoption of electrical household appliances is daily becoming more widespread, here adding a utility, and there an ornament, until in the near future we may anticipate a period when its presence in the household will be indispensable.'[3] Kennelly, as one of Edison's men, was being more than a little over optimistic in his estimation of the extent to which electricity was already present in American homes, but it was certainly clear that domestic electrification was indeed the way of the future. In fact, it was precisely because electricity offered such an alluring image of the future that many people both in Europe and America were so keen to welcome it into their homes. Installing the electric light was a tangible step into the future for middle-class families. It was a very visible expression of current prosperity and of faith in the ideal of progress.

The Statue of Liberty was adorned with electricity as well. The *Electrical Review* reported that when it was completed the 'torch of the Statue of Liberty will contain five electric lamps of 80,000 candle power, the light of which will be thrown upwards. It is believed that the light will so illuminate passing clouds that they will be visible at a distance of 100 miles. Four electric lights of 6,000 candle power will be placed at the foot of the statue, so as illuminate it. The diadem on the head of the figure will contain incandescent lamps to give the effect of jewels.'[4] The Statue of Liberty's illumination was part of a popular trend for lighting up statues and prominent public buildings with electric light. Go-ahead hoteliers and proprietors of department stores similarly turned to electrical illumination as a way of attracting and pleasing customers – and demonstrating their own commitment to a progressive future.[5]

If Lady Liberty could wear electrical jewellery, then ladies of fashion could do so too. In New York they might purchase Gustav Trouvé's electrical jewellery, imported from Paris and illustrated in the *Electrical Review* in 1885.[6] Small batteries, discreetly hidden in pockets or beneath clothing, powered tiny coloured electric lights on brooches, necklaces and hairpins. Some women went to even greater extremes to adorn themselves with electricity. As far away as Brookings, South Dakota, the wife of the manager of the local electricity company appeared at a carnival wearing 'a crown of incandescent lamps and her dress was decorated with the same ornaments'. Wires connected her jewellery to her shoes so that they lit up when she stood on some copper plates attached to a dynamo. In Iowa, a young woman appeared as the Statue of Liberty, 'her upraised right hand grasping a torch, capped with a cluster of lamps, which alone would have been sufficient to illuminate the entire room'.[7]

This culture of extravagant electrical spectacle extended to the halls of science as well, of course, and Hertz's discovery of electromagnetic waves in 1888 added another dimension to its possibilities. The English physicist Oliver Lodge was himself one of James Clerk Maxwell's avid disciples as well as an accomplished producer of electrical spectacle. It did not take long for him to make an exhibition out of the new electromagnetic waves. He recalled in his autobiography how 'I exhibited many of the Leyden Jar experiments both to the Royal Institution and the Society of Telegraph Engineers, in a lecture on "The Discharge of a Leyden Jar" where were shown many striking experiments. The walls of the lecture-theatre, which were metallically coated, flashed and sparked, in sympathy with the waves which were being emitted by the oscillations on the lecture-table – an incident which must be remembered by many of those present. This was a novel result, surprising to myself also, and I hailed it as an illustration or demonstration of the Hertz waves.'[8]

There was nothing new about this confluence of electricity and spectacle, of course. Electrical experiment and exhibition had been inseparable since the beginning of the eighteenth century. Michael Faraday had been as famous to his contemporaries as a public performer as he was as an electrical experimenter.[9] There was a longstanding recognition of an intimate connection between electricity and the body as well. Experimenters such as Giovanni Aldini had publicly electrified corpses at the beginning of the nineteenth century – providing material for satire by, among others, Edgar Allan Poe, who poked fun at the possibilities of electrical reanimation in 'Some Words with a Mummy'.[10] On both sides of the Atlantic there was by 1890 a thriving trade in electric belts and corsets as a cure for nervous disorders of all kinds. Indeed, electricity was often touted as

an antidote for the kind of nervous diseases that fast-moving industrial culture was supposed to generate. We have already seen how
George Beard, advocate of electrotherapy and specialist in such nervous diseases, had suggested in *American Nervousness* that America
was particularly prone to such disorders. They were a symptom of
the speed with which the country was hurtling into the future.[11]

And so, at the beginning of the 1890s, Tesla was set to put his
own electrified body on show to demonstrate his vision of the electrical future. Living in New York at the end of the 1880s he cannot
have escaped recognizing the confluence of electricity and exhibition. The evidence was all around him. Electricity seemed made for
spectacle. Spectacle, after all, was the chief weapon that Edison had
employed in his efforts to discredit George Westinghouse's alternating current systems. Newly returned from the Paris Universal
Exposition where he had witnessed at first hand the way in which
Edison used exhibition as a way of promoting his own inventions,
Tesla recognized that he had to turn himself into a showman as well.
With the electric chair making the news, and Harold Brown's electrical experiments with the city's stray cats and dogs drawing curious
audiences, Tesla may well have concluded too, that making himself into an electrical spectacle was the key to making a name for
himself as someone to be reckoned with in the world of invention.

What Tesla now needed was a public platform for himself
and his inventions. In search of that platform he turned to the
American Institute of Electrical Engineers once more, conscious of
the boost his first performance there had given his reputation. As
luck would have it, William Anthony, who had helped Tesla gain a
platform at the Institute in 1888, was its president in 1891. Another
key Tesla ally, Thomas Commerford Martin, chaired the committee on papers and meetings. With their help, Tesla's place on the
schedule was secured and his performance took place in the lecture

hall at Columbia College in New York on 20 May 1891. Tesla had prepared as carefully for the show as would any stage conjuror. He set up his high-frequency alternator, powered by an electric motor, in the college workshop, with cables leading to the oscillating transformer on stage in the lecture hall. He could control the speed of the alternator, and hence the frequency of the oscillating transformer, with a convenient switch.

'Of all the forms of nature's immeasurable, all-pervading energy, which ever and ever changing and moving, like a soul animates the inert universe,' he told his audience of electrical engineers, 'electricity and magnetism are perhaps the most fascinating.' And Tesla certainly succeeded in fascinating them.[12] He introduced his audience to his oscillating transformer and illustrated the amazing effects he could produce. He demonstrated sparks and discharges of various kinds, showing how they varied with frequency and potential – warning that in some instances it might be 'inadvisable' to put oneself in the way of the discharge, and showing that in others it might be done with impunity. The main point of the performance, however, was to demonstrate the feasibility of electric light without wires. Tesla suspended two metal sheets from the ceiling, connected to his oscillating transformer. When he stepped between them, holding a discharge tube in each hand, the tubes began to glow under the influence. This really was the 'ideal way of lighting a hall or room' because it meant that 'an illuminating device could be moved and put anywhere, and that it is lighted, no matter where it is put and without being electrically connected to anything.'[13]

The whole thing was a huge success. The magazine *Electricity* enthused that 'no paper for months has excited such wide spread interest as that of Nikola Tesla, read before the American Institute of Electrical Engineers. His experiments with alternate currents are wonderfully interesting and suggestive. All attempts at securing light without heat must excite general interest, even though the investigation, as in parts of Mr. Tesla's work, is in the most occult domain of electrical theory.'[14] That hint of magic was something Tesla himself would do nothing to discourage. The journalist Joseph Wetzler, writing in the popular *Harper's Weekly*, thought that the lecture had finally cemented Tesla's reputation. In 'one bound he placed himself abreast of such men as Edison, Brush, Elihu Thomson, and Alexander Graham Bell'. The possibilities of wireless light seemed endless: 'suffice it to say that without wires or pipes of any kind to hamper the artist or decorator, effects may be produced which will bring fairy-land within our homes.'[15]

Having conquered the New World, it was time for Tesla to conquer the Old World too. London and Paris were still the twin centres of the electrical universe and if Tesla really wanted to make his mark he had to triumph there as well. He had financial interests to protect too. His claims to the invention of the polyphase motor might be recognized in America, but his rights in Europe were on rather shakier grounds. He needed to enforce his rights and make sure that the patents he had taken out in several European countries were properly protected. Tesla therefore boarded the Royal Mail Ship *Umbria* for the ten-day journey from New York to Liverpool on 16 January 1892. Owned by the Cunard line, the *Umbria* was one of the fastest, largest, and most luxurious of the ocean liners. The ship was capable of travelling at 19 knots, and had won the coveted Blue Riband just a few years previously.

On his arrival in London, Tesla was honoured with an invitation to stay at the house of William Henry Preece. Born in Caernarfon, Preece was a thoroughly self-made man, who had worked his way up through the ranks of telegraphic engineering to become the Post Office's chief telegraph engineer.[16] His support was going to be vital for Tesla's efforts to sell himself to this audience of hard-nosed

British electricians. Just as important would be the support of William Crookes, president of the recently renamed Institution of Electrical Engineers. With Crookes's support, Tesla had already secured an invitation to lecture before that body, and attended its annual general meeting on 28 January.[17] Not only that, the lecture would take place at the Royal Institution, in the very lecture theatre where Michael Faraday and John Tyndall had performed. It would be repeated the following night before the members of the Royal Institution. Tesla would be talking about and showing off his oscillating transformer from the podium where Faraday had first announced and demonstrated electromagnetic induction.

Tesla introduced his lecture with a nod towards his new patron. He recalled his first reading, many years previously, of Crookes's work on radiant matter, suggesting that the 'fascinating little book' was the real origin of all his later discoveries.[18] Tesla proceeded to belabour his audience with spectacle upon spectacle. 'Here,' he showed them, 'is a simple glass tube from which the air has been partially exhausted. I take hold of it; I bring my body in contact with a wire conveying alternating currents of high potential, and the tube in my hand is brilliantly lighted.' 'Here,' he showed them again, 'is an exhausted bulb suspended from a single wire. Standing on an insulated support, I grasp it, and a platinum button mounted on it is brought to vivid incandescence.' Again, 'as I stand on this platform, I bring my body in contact with one of the terminals of the secondary of this induction coil – with the end of a wire many miles long – and you see streams of light break forth from its distant end, which is set in violent vibration.'[19] And so the relentless exhibitionism continued, with Tesla quite literally the centre of attention.

The lecture was a sensation. The *Standard* called it 'marvellous and astonishing'. Its correspondent marvelled in particular at 'the safety with which the lecturer appeared to handle' the tremendous currents. 'It was startling to see,' the report went on, 'the loud snapping streaks of bright lightning eight or ten inches in length flashing off from the induction coil, in pressure equal, it was understood, to 150,000 volts, the offtake of the same current passing through the body of the lecturer to a flaming phosphorescent sword, as it were, formed by a two-foot glass tube.'[20] *The Times* not only reported on the lecture delivered by the 'young but distinguished electrician from America',[21] but even devoted an editorial to the performance and its significance. Tesla had 'for two hours held a professional audience entranced at the Royal Institution'. His lecture had offered 'a stimulating expansion of our speculative ideas' and penetrated into the 'borderland where light, heat, electricity, chemical affinity, and forms of energy which we cannot confidently identify with any of these, meet and blend'.[22]

Tesla was feted and celebrated. He joined the Lord Mayor's entourage on an inspection of the Crystal Palace's recently opened Electrical Exhibition. He was complimented in the speeches after dinner and gave a little speech of his own.[23] The Exhibition's advertisements were soon promising that visitors would see 'Mr. Tesla's astounding experiments repeated daily'.[24] Tesla himself was on his way to Paris, however, where he gave a lecture to the Société de Physique and the Société International des Electriciens on 19 February. There he repeated many of the experiments that had caused so much excitement at the Royal Institution – and the Parisian electricians were just as impressed as those in London had been. 'The French papers this week are full of Mr. Tesla and his brilliant experiments,' reported the *Electrical Engineer*. 'No man of our age,' the report said, 'has achieved such a universal scientific reputation in a single stride as this gifted young electrical engineer.'[25]

To some of those 'who had the good fortune to be present at Mr. Tesla's lecture', the 'young scientist' seemed 'almost as a

Tesla lecturing before the Société de Physique during his trip
to Europe in 1892. (*Scientific American*, 26 March 1892)

prophet'. It even seemed that 'the dream of the inventor is broader
and his views more exalted than the experiments that he presented
to us allow to be seen.' Tesla's ambition, the electrician Édouard
Hospitalier thought, was 'to transform the energy of the medium
that environs us, and which is very evident by its numerous mani-
festations, into light, or at least to obtain therefrom radiations of
the same wave length and same frequency as those that produce
luminous sensations'. The audience at his lecture were 'witnesses of
the dawn of a nearby revolution in the present processes of illumination'.[26] Sadly for Tesla though, he had little time to bask in the
adulation. While in Paris he received news that his mother was
dying, and he rushed back to Gospić to be with her. On his way
back, he visited Hertz at his Bonn laboratory, and was bemused
when the experimenter refused to accept his explanation of what
the Hertzian waves really demonstrated.[27]

The Tesla that returned to New York at the end of his European adventure was in many ways a different man from the one who had boarded the *Umbria* several months earlier. His return merited a brief notice in the *Electrical Review*: 'Now that Mr. Nikola Tesla has returned from Europe we may expect to hear of greater and further advancements in his work with alternating currents of high frequency,' the journal predicted.[28] It had been not only a triumph of showmanship, but a lucrative business trip as well. The *Electrical Engineer* reported that 'the reports current in Europe as to the control of Tesla motors secured there by leading concerns are well founded.'[29] Tesla was no longer an inventor on the make. He was a fully self-made man, recognized as someone who would have a central role to play in the future of electricity. His European celebrity, and the important allies he had secured on the other side of the Atlantic, all contributed to his standing as an American inventor of the future. It was his success at showmanship that had proved decisive in providing him with his chance to remake the electrical landscape.

Tesla was certainly clear in his own mind about the future that lay ahead of him. Hospitalier's remark in his account of the Paris performance about the young electrician's high ambitions was both perceptive and prescient. Returning to America, Tesla no longer (if he ever had) regarded himself as a mere inventor of dynamos and motors. He was going to remake the electrical future. It came to him, or so he recalled in his memoirs, during a thunderstorm in the mountains near Gospić. 'Here was a stupendous possibility of achievement,' he thought. 'If we could produce electric effects of the required quality, this whole planet and the conditions of existence on it could be transformed.' To succeed in this aim he would need to find ways 'to develop electric forces of the order of those in nature'.[30] Back in New York, Tesla was already thinking about ways of turning that ambition into a reality.

CHAPTER 10

Harnessing Nature

L ooking back in his memoirs, Tesla recalled that as a child he had fantasized about Niagara: 'I was fascinated by a description of Niagara Falls I had perused, and pictured in my imagination a big wheel run by the Falls.' He was determined to 'go to America and carry out this scheme'.[1] Writing about Tesla's London performance in the *Nineteenth Century*, the electrical engineer James Edward Henry Gordon had clearly been thinking about Niagara as well. 'On the same table, on which Mr. Tesla's experiments were shown a few days ago, there swung,' he reminded his readers, 'in the year 1834, a delicately balanced galvanometer needle, under the influence of the first induction current, produced by the genius of Faraday.' That force, too, had been small at the beginning, 'probably not greater than the forces lighting Mr. Tesla's tubes, yet that force has now developed one of the great industries of the world'.[2]

That force now powered 'millions of lamps in London and elsewhere, in America it drives cars on thousands of miles of railways, and will soon distribute the power of Niagara Falls to the inhabitants of neighbouring States'. Gordon speculated that Tesla's discoveries would 'some day harness to our machinery the natural forces, which from the beginning of time have literally been slipping through our fingers'. He imagined that if Tesla's dreams became reality, 'we shall see a social and political change at least as important as that caused by the railway system or the electric telegraph.' It would be a world where 'manual labour will become unnecessary, as unlimited power will be available at every man's hand.' Niagara was indeed on Tesla's mind as he travelled back to America. Plans were already under way to turn the Falls' energy into practical power, and Tesla recognized the opportunity those plans offered to demonstrate the potential of his system too.

Tesla was certainly not the only one who had been dreaming of ways to turn the Falls' raw power into useful work. In 1876 William Siemens had visited the United States to attend the Centennial Exhibition in Philadelphia. He also visited the Niagara Falls. As his biographer put it, the 'stupendous rush of waters filled him with fear and admiration, as it does everyone who comes within the sound of its mighty roar'. But, Siemens being Siemens, 'his scientific mind could not help viewing it as an inexpressibly grand manifestation of mechanical energy.' For Siemens, electricity was clearly the answer to the problem of how to make that energy available and useful. He knew that the 'dynamo-machine had just then been brought to perfection, partly by his own labours; and he asked himself, Why should not this colossal power actuate a colossal series of dynamos, whose conducting wires might transmit its activity to places miles away?'[3]

A year later, Siemens fleshed out his vision of Niagara's future in his presidential address to the Iron and Steel Institute. There he talked about the possibilities of water power, and 'the magnitude of power which is now for the most part lost, but which may be, sooner or later, called into requisition'. Niagara was just so much wasted energy. Siemens calculated that 'the force represented by the principal falls alone amounts to 16,800,000 horse-power … all the coal raised throughout the world would barely suffice to pro-duce the amount of power that continually runs to waste at this one great fall.' If that 'water-power be employed to give motion to a dynamo-electrical machine, a very powerful electric current will be the result'. Siemens reckoned that a 'copper rod three inches in diameter would be capable of transmitting 1000 horse-power a distance of say thirty miles, an amount sufficient to supply one quarter of a million candle-power, which would be sufficient to illuminate a moderately sized town'.[4] In 1881, just a few years after Siemens's prophecy, a small hydro-electric power station was indeed built at the Falls by the Niagara Falls Hydraulic Power and Manufacturing Company.

Another very eminent advocate of the Falls' potential as a source of huge quantities of electric power was Sir William Thomson – one of the leading figures in the worlds of electrical theory and electrical engineering alike. As early as 1879 he, like Siemens, had carried out some back-of-the-envelope calculations on the economics of transmitting electricity from Niagara while preparing evidence for the British House of Commons Select Committee on Electric Light. He presented the results of further calculations to the British Association for the Advancement of Science in 1881.[5] A decade or so later, Thomson was appointed chairman of the International Niagara Commission established to consider the best way of exploiting the Falls for power generation. The Commission had been established by the Cataract Construction Company and its president Edward Dean Adams. Adams was convinced that the best way of profiting from Niagara was by finding ways to transmit the electric power generated there to more distant cities like Buffalo, rather than using it locally.

Up until this point, George Westinghouse had shown relatively little interest in the prospects for Niagara. This was partly as a result of the financial difficulties suffered by his companies at the beginning of the 1890s, but also because he remained unconvinced about the economics of producing and transmitting electricity at Niagara. Accordingly, when Adams at the end of 1891 approached Westinghouse as well as a number of other companies, including

Edison's, to bid for contracts, he did nothing. Edison on the other hand proposed a plan that would use a direct current system to provide electricity for local use and an alternating current system to transmit electricity to customers further afield. It was not until the end of 1892 (when Tesla was back in America) that Westinghouse started bidding for the Niagara contract. Key to his conversion were the improvements his engineers had carried out on Tesla's polyphase motors and their development of ways to make it easier to convert polyphase alternating current so that potential customers could use it to power existing equipment.

As far as Tesla was concerned, the plans to generate electricity at Niagara offered an unparalleled opportunity to demonstrate the power of his vision for an electrical future. Recalling events a quarter of a century or so later, as he was awarded the Edison Medal in 1917, Tesla was still sure that his own role in the project had been decisive. It was he who had persuaded Adams that polyphase alternating current was the solution to his problems. 'Mr. Adams was much impressed with what I told him,' he remembered. The two of them 'had some correspondence afterwards, and whether it was in consequence of my enlightening him on the situation, or owing to some other influence, my system was adopted. Since that time, of course, new men, new interests have come in, and what has been done I do not know, except that the Niagara Falls enterprise was the real starting impulse in the great movement inaugurated for the transmission and transformation of energy on a huge scale.'[6]

That was the scale of Tesla's ambition. The prospect of limitless (or at least practically limitless) energy was something that he would return to many times over the coming decade. In his correspondence with Adams, Tesla was adamant that only his system and his improved polyphase motors would deliver on that ambition. His latest designs had 'never been criticized by the adversaries of my system and for a good reason, because it is the most efficient form of electric machine that has been produced to this day'. He told Adams that 'in such machines under favourable conditions an efficiency of 97% can be obtained.'[7] It was a bold claim – and typical of Tesla's increasing faith in himself and his inventions. He brushed aside Adams's worries about the patent situation and efforts by General Electric (the result of the recent merger between the Edison Electric Company and Thomson-Houston) to develop their own rival polyphaser system. Nothing could seriously rival his, he told him.

There remained rival direct current competitors to see off as well. William Thomson, recently elevated to the House of Lords as Baron Kelvin of Largs, had been a supporter of direct current systems since those back-of-the envelope calculations that he had carried out a decade or so earlier. When Adams told Tesla that a 'prominent advocate' was pushing for a direct current alternative it seems likely that Thomson was the culprit (and that Tesla was well aware of his identity). There could be no question, Tesla reminded Adams, that his system was 'simpler, cheaper, and more efficient in general'. It enabled 'absolutely constant speed, facility of insulating for high voltage', and 'easy conversion to any voltage'. Those advantages were 'practically unattainable in the direct system, especially when transmissions at a great distance are contemplated'. Of course, a direct current system was possible in principle, but 'the question here is to achieve a practical commercial success and the best and safest appliances should be employed by all means.'[8] Those were his, of course.

Tesla was never involved in the practical work of turning the plans for generating electricity at Niagara into a reality. He was no longer employed by Westinghouse, after all. Despite Tesla's ambitious persuasions, Adams and the Cataract Construction Company took their time in deciding what system they were going to deploy at Niagara Falls. It was May 1893 before they announced that they would adopting a two-phase alternating current system there. A few months later, in October 1893, Westinghouse was awarded the contract to build the generators. By the end, the Niagara Falls powerhouse would contain ten of these Westinghouse generators, designed originally by the British electrical engineer George Forbes, who had been one of the members of the International Niagara Commission and would be the consultant engineer for the works. Each generator was capable of producing up to 5,000 horsepower. The whole thing was a huge engineering challenge. Not just the generators and the powerhouse needed to be built, but a network of canals and tunnels that would channel the waters of Niagara to where their flow could be made useful.

For many visitors to Niagara Falls, these gargantuan engineering works offered an added attraction. As Coleman Sellers expressed it in the tourist guide *The Book of Niagara*, for 'the first time in the history of Niagara Falls, attractions other than those furnished by nature are offered; not only to the mere pleasure-seeker, but to the scientific world generally, in the attempt that is now being made to utilize some small portion of the power of the great cataract on a scale that is vast compared to all previous attempts at such utilization'. Rather than detracting from the Falls' grandeur, the works contributed to it, by 'adding to what would otherwise attract the visitor to the place, the visible progress of a gigantic engineering enterprise that has no precedent in the civilized world or that can be compared to any of the similar works of man in the results that may flow from the venture'.[9]

Forbes, as the consulting engineer, was certainly proud of what had been achieved. Looking down from the pinnacle of a 'small Eiffel tower' erected at the site, he described to readers of *The Times* how he could 'see a new world created'. There was a 'wide canal leading water from the river into that gigantic tower-house where three turbines are set up to drive three dynamos of 5,000 horse-power each'. From there, the water 'passes through a 7,000-feet tunnel under the town, emerging below the Falls, the tunnel being capable of developing 100,000 horse-power'. In the distance was 'the model village for working men, and improved sewage-works with drainage, pumps for water supply, electric light, and well-paved streets'. It was a vision of a future in which 'the power could be sent much more than a hundred miles, and still be more economical than steam, even though coal is cheap there.'[10]

For some, Niagara seemed to hold out the prospect of practically infinite power. To all intents and purposes, a 'reliable, uniform water supply, such as is ideally represented at Niagara Falls, is the nearest obtainable approach to perpetual motion'.[11] They fantasized about a future in which more and more of America would be powered by its waters. At that moment, profitable transmission might only be possible up to a distance of 150 miles, but '150 miles from Niagara Falls in a straight line brings us to within ninety miles of the city of New York.' Before long, its reach would embrace 'Columbus,

Niagara Falls. (*Scientific American*, 11 November 1899)

Ohio, touching Washington, D.C., including Philadelphia and New York, and the whole of the states of Pennsylvania, New York, part of Maryland, the northern part of Virginia, and West Virginia, more than two-thirds of Ohio, fully three-quarters of Michigan, beside reaching to Montreal in Canada'.[12] Chicago and New England would soon be drawing their electricity from Niagara too.

Niagara occupied a prominent place in the American imagination. Mark Twain placed the Garden of Eden there. Another commentator invoked the 'sense of power and of mystery' there, 'which overcame the mind'.[13] 'Spectators by the million' had unconsciously 'revealed something of themselves in various efforts to disclose to others the essential character of the Falls of Niagara, confessedly incomparable with any other natural object.' This was nature at its wildest, sublimest and most dangerous – so what better expression of American ambition than to make it into a servant. Niagara radiated a sense of power, and it was 'through this very sense of resistless power that the falls speaks to minds of great dignity and self-restraint'.[14] Even Lord Kelvin could wax lyrical about how he looked forward to 'the time when the whole water from Lake Erie will find its way to the lower level of Lake Ontario through machinery, doing more good to the world than even that great benefit which we now possess in contemplation of the splendid scene which we have before us in the waterfall of Niagara'.[15]

The machinery that tamed Niagara offered a potent symbol of America's future power. The powerhouse itself, designed by the architect Stanford White, expressed 'in its unique and impressive architecture the enormous power to be generated within, and giving in itself some hint of the capacity of man to master the forces of nature'. In the past, thousands of people might 'have looked upon the Niagara cataract with emotions such as are inspired by no other natural object', but in the future 'when the visitor who goes to look upon the sublime cataract discerns how men have at last found a way to take from it some portion of its force, and how the whirl of the wheels of factories, the illumination of cities, the steady motion of vessels upon the artificial waterways are all but responsive throbbings to the thunder of the cataract, perhaps he will consider that a yet finer epic has been written by the energy of those who have been engaged in subduing the forces of nature that they may serve mankind.'[16]

Tesla might not have had any direct involvement in the actual construction of the Niagara project, but it still contributed to his growing reputation. As far as many newspapers were concerned, this was Tesla's scheme – and Tesla did little to disabuse them of the notion. The *Buffalo News and Sunday Express* led their Niagara coverage with a large photograph of Tesla. 'No name is more brilliantly associated with recent progress in electricity than that of Tesla,' Madison Buell told his readers. His 'new thoughts and his experiments have developed and corrected all former notions of the science – given it a new character from that ever before conceived'.[17] *McClure's Magazine* was emphatic that it was only when 'Nicola [*sic*] Tesla, that illuminating wizard of electricity,' stepped up to the mark that investors were persuaded to finance such as ambitious project.[18] Even in the electrical press, discussions of the ongoing work at Niagara emphasized that this was 'the Tesla multiphase system', or 'the Tesla polyphase'.[19]

For the *New York Times*, to 'Tesla belongs the undisputed honor of being the man whose work made this Niagara enterprise possible'.[20] Tesla, in fact, did not set foot on Niagara's shores until the work was practically finished. Even that, though, was food for the newspapers. The *Western Electrician* offered breathless reportage, describing how for 'four years he had refused to leave his work and visit the Falls, preferring to work out his theories and await the proper time to see them put into practical use and operation'. Now, he was 'delighted with the manner in which his discoveries had been adapted to practical use by the engineers, and he unhesitatingly declared that there was no doubt of the success of the gigantic undertaking and that power would be transmitted to Buffalo without the least semblance of failure in any important detail'. Niagara represented the beginning of a future without steam. 'The time will come when steam will not be used for commercial purposes,' he told reporters.[21]

In a spectacular coup, Thomas Commerford Martin, one of Tesla's oldest supporters, contrived to have some of Niagara's output transmitted all the way to New York in May 1896 for the National Electrical Exposition. *Scientific American* reported that a 'model of Niagara River, the power house, the town and the discharge tunnel will be exhibited', and the 'turbines will be run for a time each evening with electricity generated at Niagara Falls and transmitted to New York by two copper wires of the Western Union Telegraph Company'. Visitors would be able to listen to the sound of the falls over the telephone. There were even rumours that Niagara electricity might be transmitted to Europe using the Atlantic cable.[22] This was Tesla's future in microcosm. There was some controversy as to whether or not the electricity running the model really came from Niagara.[23] But as a piece of showmanship it made very explicit the role that exhibition played in the process of making the future. Even the controversy that surrounded it demonstrated the blurred lines between fact and fiction in these displays of inventive power and ingenuity.

CHAPTER 11

The Greatest
Show on Earth

The contract to build the generators for Niagara was not
George Westinghouse's only triumph of the early 1890s.
Westinghouse and his Electrical Company were also awarded the
contract to provide the electricity, as well as the necessary machinery and equipment, for the World's Columbian Exposition that
would open in Chicago in 1893. Recovering as the company was
from financial insecurity, this was a major coup. It was also part of
a concerted effort by Westinghouse to demonstrate once again the
decided superiority of his alternating current system. Even more
than the acquisition of the Niagara contract, it demonstrated to
the American public that the future lay with alternating current – a
system almost entirely identified by them as originating with Tesla.
Westinghouse had won the contract by under-bidding his competitors by a substantial margin. To make sure that the company could
deliver on its promises without massive losses, its engineers had to
find new and more economical ways of working.

To supply the exhibition, they 'devised and constructed in
less than six months larger machines than had ever been built for
this work before, and on radically different lines, embodying the
principles of the alternating system of transmission'. As a result,
'hundreds of thousands of dollars' worth of copper wire were saved,
as it was possible to send the current under high pressure to its
destination on small wires, and then transform it down at the point
of utility.'[1] Westinghouse and his engineers also needed to develop

a new type of incandescent light, since Edison, having failed in his own attempt to win the contract, would not allow them to use his patented design. Electricity was going to dominate the exhibition, and it played a key role in its preparation as well. It 'helped to prepare the material, to hoist the heavy beams and trusses, to paint the buildings, and at the same time to prolong the labors of the overworked engineer and mechanic, and light the rough or muddy pathway of the Columbian Guard'.[2]

The Columbian Exposition had taken years of planning. Discussions about how to mark the 400th anniversary of Christopher Columbus's arrival in the New World had been going on since the Centennial Exhibition in Philadelphia in 1876, celebrating the 100th anniversary of the declaration of independence. By the late 1880s there was fierce competition between Chicago, New York, St Louis and Washington for the opportunity to host the exhibition. Chicago won the battle by defining itself as the city of the future. Holding the exhibition there would 'display to the world an American wonder, a city of a million and a quarter of inhabitants, and but half a century old'. The act 'to provide for celebrating the 400th anniversary of the discovery of America by Christopher Columbus, by holding an International Exhibition of arts, industries, manufactures and the products of the soil, mine and sea, in the City of Chicago, in the State of Illinois' was signed into law by the President on 25 April 1890.[3]

The Exposition was opened with spectacular ceremony on 1 May 1893. A 'hundred thousand persons were crowded into the Plaza, eager to see all that was to happen'. Electricity was at the heart of it. Once the President had finished his speech, 'he touched an electric button on a table before him.' As soon as he did so, 'there was a flash of color from a thousand staffs crowning the great buildings' as 'flags of every nation' were hoisted. With the flags flying, 'bands began to play, steam whistles to blow, vessels in the harbor to fire salutes from their guns, and from the mighty throng went up that grandest of songs that ever rises from earth to heaven, the cheers of a multitude for a work that is grand and good.' Immediately, 'every wheel of all the great machines began to turn as if by magic.' It was all a 'glorious triumph of man's effort, and the payment for toil and anxiety and rebuff'.[4]

In some ways, the most striking feature of the Columbian Exposition was its sheer scale. Its grounds covered 630 acres along the shores of Lake Michigan. The grounds, designed by the Boston architect Frederick Law Olmsted, were laid out around a network of canals and lagoons, with the Grand Basin at its centre. At one end of the Basin there stood a 65-foot-tall golden statue of the Republic. There were almost 200 buildings arranged around the grounds, including fourteen main buildings. The largest of these buildings, the Manufactures and Liberal Arts Building, overlooking the Grand Basin, was enormous. It was almost 1,700 feet long and 800 feet wide. In all it covered more than 30 acres of ground. Its gargantuan size was designed to be the Exposition's pièce de résistance – the American answer to the Eiffel Tower at the Paris Universal Exposition a few years earlier. Like all the other buildings, it was faced in white, giving the Exposition its nickname of the White City.

As well as the Manufactures Building, there were buildings devoted to Agriculture, Electricity, Fisheries, Forestry, Horticulture, Machinery, Mines and Mining, and Transportation. There was a central Administration Building as well as a Women's Building and an Anthropology Building too. The Exposition even had its own railway station. Steam ships could dock at the pier that extended out onto the lake. Another innovation was a moving walkway along the shore. One major attraction was the Ferris Wheel. Designed by

George Washington Gale Ferris, the wheel was 264 feet high and carried 36 cars that could each hold up to 60 people. The whole thing was powered by steam engines generating 1,000 horsepower. In addition to the main buildings, there were buildings to house exhibits from individual nations, as well as exhibits from the various states of the union. While not an official part of the Exposition, Buffalo Bill Cody brought his Wild West Show to Chicago and set up his camp on the fringes to take advantage of the millions of visitors who flocked to the White City to see the future.

And visitors did indeed flock there in their millions. During the six months for which the Exposition lasted, something like 27 million people passed through. Reactions were almost uniformly positive. Candace Wheeler in *Harper's Magazine* called the Exposition 'a shining vision, serenely awaiting the admiration of the world'.[5] Indeed, she concluded, 'the beauty of the Dream City' was in fact 'beyond even the unearthly glamour of a dream'.[6] The *Century Magazine* warned its readers that 'no one can see the whole of a Fair like this, inside and out; and time, energy, and disappointment will be saved if a plan of campaign is prepared in advance, and the mind is trained to feel that it must be followed.' 'Most Americans,' the *Century* reporter thought, would 'go to the Fair with some serious purpose before them – if not to study carefully any of the collections, then to make a careful general survey of the Fair itself, as illustrating the present condition of our nation from many points of view, and likewise its promises and prospects for the future.'[7]

As the exhibition drew near its end in October, *Harper's Weekly* summed up the reaction. Everyone should remember, it said, how 'the genius of the country has created a work of surpassing grandeur which should not be permitted to pass away without having exerted to the widest extent its enlightening and elevating influences upon the living generation'. This was a show that would not be forgotten.

It had offered the ordinary American citizen 'a more just conception of what he is, and a more wholesome appreciation of what it should still be his ambition to become'. It had been 'a gorgeous dream of human genius' that would 'long be spoken of by this and coming generations as one of the greatest marvels of the closing nineteenth century'. The magazine encouraged everyone who had not already been to go at once. They 'should not fail to bestow upon their children the boon of the enlightening and ennobling impressions which this grand spectacle conveys'.[8]

At the very heart of this extravaganza of the future was electricity. The Electricity Building stood on the other side of the North Canal from the monumental Manufactures Building, alongside the Mines Building. Walking past Benjamin Franklin's statue outside the main southern entrance, visitors would find themselves in the Bell Telephone Company exhibit. They could listen to opera transmitted from New York, or watch the switchboard, and the 'Hello!' girls operating it, that connected all the telephones in the exhibition grounds. The Western Electric Company had chosen an Egyptian theme. Visitors could watch 'a group of Egyptian maidens, of the time of Ramses the Second, operating a telephone board', or 'men of the same period laying telegraph lines'. The whole thing was a commentary on past, present, and future – and 'the conceit', it was said, 'is very popular'.[9] Another exhibit offered a 'tower covered with lamps, from the top of which are made to shoot in four directions long streaks resembling forked lightning'.[10]

The Brush Company exhibit featured 'a full line of central station apparatus and railway work', as well as a display of electric light. The lights themselves were out of sight, and 'focus their rays upon the ceiling, which is a dome, tinted cream colour'. This was 'by far the best piece of lighting of its character in the building'.[11] The Fort Wayne Electric Company also featured a 'fine exhibit' of 'a commercial lighting station in full operation', that 'shows to the public exactly what should go into a regular station to meet any and all demands of service'.[12] H.W. Axford had on show an 'incubator for hatching poultry'. In this incubator, not 'only does electricity replace the brooding hen, but cares for the motherless chick after the process is completed, and does it in a manner that is precision itself'. It even completed the business in less time than would be required by an actual living hen.[13]

Westinghouse, of course, had a very substantial presence in the Electricity Building. Their exhibit featured 'a mural decoration in incandescent lamps, showing the figure of Columbus with the names, dates, 1492–1892, and some beautiful scroll work'. This installation needed '1988 incandescent lamps of 16 candle power in frosted and plain white and colors'.[14] General Electric and Edison had a huge display as well, including 'lamps ranging from a power of ¼ of a candle to 250 candle-power, and examples of all his lamps from the very first to the latest are shown, as well as of the materials for and the various stages of their manufacture'.[15] Edison was clearly very keen to emphasize the part that he had played in generating this new electrical age, as well as the role he and his companies hoped to occupy in dominating its future. General Electric's key exhibit, though, was the 'Tower of Light' that was 'intended as a glorification of the Edison lamp and the Edison system of incandescent lighting', and would 'linger in the memory of the visitor as one of the beautiful spectacles at the Fair'.[16]

Tesla had his very own show as part of the Westinghouse exhibit. The apparatus on show represented 'the results of work and thought covering ten years'. It included a ring generating a powerful magnetic field, which 'exhibited striking effects, revolving copper balls and eggs and bodies of various shapes at considerable distances and at great speeds'.[17] On the ground floor was 'a special dark

The Electrical Building at the Columbian Exposition.

building' inside which were a variety of exhibits 'to illuminate the recent and absorbingly interesting developments made by Nicola [*sic*] Tesla, of the use of high tension alternating currents'. These included 'two large plates placed at approximately eighteen feet from each other'. Between them were 'two long tables with all sorts of phosphorescent bulbs and tubes; many of these were prepared with great care and marked legibly with the names which would shine with phosphorescent glow'. The names included 'Helmholtz, Faraday, Maxwell, Henry, Franklin, etc.'.[18] Tesla was leaving very little doubt about the place that he ought to occupy in the electrical pantheon.

Edison and his Tower of Light might have been triumphant in the struggle for spectacle inside the Electrical Building, but outside it was all Westinghouse. 'The Westinghouse Company also has, as an exhibit, almost the entire display of incandescent lighting on the grounds,' commentators noted. That in itself had been a gargantuan effort – 'they have built and installed, within the year, twelve generators of a total capacity each of 15,000 incandescent lights of 16 candle-power each.'[19] The power was all controlled from a 'great white marble switchboard' that was 'designed for 12 two-phase and two single-phase alternators, four exciters and 40 circuits and controls nearly all the incandescent and a large number of arc lights in the grounds and buildings including the Midway Plaisance'.[20] Visitors to the exhibition could see all these works on display in the Machinery Building. It was a hugely impressive show of labour and organization, as well as a spectacle in its own right.

Visitors would have seen electricity everywhere they looked. The Columbian Exposition was 'a magnificent triumph of the age of Electricity'. Apart from some of the machinery in the Machinery Building, 'all the exhibits in all the buildings are operated by electrical transmission.' In the grounds of the exhibition themselves, the 'Intramural Elevated Railway, the launches that ply the Lagoons, the Sliding Railway on the thousand foot pier, the great Ferris Wheel, the machinery of the Libby Glass Company on the Midway, are all operated by electrically transmitted energy'.[21] Everywhere and everything 'pulsates with quickening influence of the subtle and vivifying current'. It really did seem like the sudden and unexpected arrival of the future: 'All this hardly seems strange to the boy who cannot look behind him into even the very near past, but to those of us who remember former Expositions there appears to have been some radical revolution at work to accomplish what we now see before us.'[22]

The electric fountains were widely admired. Crowds in their thousands 'stand at points of vantage each evening to watch the ever-changing beauties of these fountains'. They were 'without a rival in ancient or modern days in hydraulic or electrical design'. The electricity powering the fountains also charged the little electric launches that sailed the canals and lagoons. Fifty of these 'beautiful little boats', each carrying up to 30 people, could each hold enough charge for ten or twelve hours' sailing.[23] Another wonder was the six-foot Schuckert electric searchlight positioned at the pinnacle of the Manufactures Building – the biggest lamp in the world on top of the biggest building. The strength of its central arc light was 57,000 candle power, increased to 194,000,000 candle power by the parabolic mirror. It was claimed that on a clear day, and from a high enough point, the light was visible as far as 100 miles away.[24]

The hyperbole of the event itself made the Exposition an ideal backdrop for Tesla's projection of himself. The International Electrical Congress met in Chicago in August, and Tesla delivered a lecture on Mechanical and Electrical Oscillators to the Congress on its final day, held in the Agricultural Building on the World's Fair grounds.[25] The following Tuesday he repeated the lecture in front of a general audience that included 'von Helmholtz and a number of other scientists who had been unable to attend the first lecture'. In the evening, 'Mr. Tesla repeated his beautiful and interesting high tension and high frequency experiments in the dark pavilion of the Westinghouse company in the electricity building.'This time the audience included W.H. Preece, Tesla's patron during his triumphant visit to London two years earlier. He repeated that performance, and 'instructed and entertained other parties of visitors at intervals during the week at the same place with similar demonstrations'. He told journalists that 'he hoped to continue his investigations until he could present the world a system of mechanical and electrical vibration of commercial utility.'[26]

The success of the Columbian Exposition – and the key role of its electrical exhibits in that success – cemented electricity in the public mind as the technology of the future. It looked as if the electrical future had already arrived. A house equipped with the contrivances on display at the Electrical Building 'would be a marvel of comfort, and would be luxurious beyond all desire'. The White City symbolized the American future, and that future was clearly going to be electric. 'Perhaps another "electrical exhibition" a decade hence would show as great an advance over the present one as it does over the Centennial,' gushed J.P. Merill in the official summing-up of the Fair's achievements. Even if that seemed impossible, 'every one will admit that the applications of electricity are still in their infancy, and the coming generations will certainly see wonderful advances in this science of "chaining the lightning and harnessing the thunderbolt".'

The difference that the past decade had made to Tesla and his public standing had been remarkable. In 1883 no one, other than Tesla himself, would have recognized him as the man of the future. He was a jobbing electrician looking for a way to ply his trade. As 1893 and the World's Columbian Exposition drew to a close, Tesla was the future. He was the man whose inventions had made Niagara the source of endless streams of electrical power. He was the power behind the White City. Writing two years later in the *Century Magazine*, Tesla's friend Thomas Commerford Martin summed up succinctly the place he now occupied in the popular imagination: 'Electricity has, indeed, taken distinctively new ground of late years; and its present state of unrest – unsurpassed, perhaps, in other regions of research – is due to recent theory and practice, blended in a striking manner in the discoveries of Mr. Nikola Tesla, who, though not altogether alone, has come to be a foremost and typical figure of the era now begun.'[27]

PART 4

SELLING
THE FUTURE

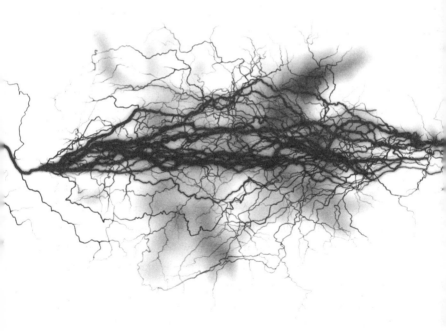

CHAPTER 12

In the Ether

Throughout the mid-1890s, while the world was marvelling at the wonders of the Chicago exhibition and the Niagara project, Nikola Tesla continued to experiment with his oscillating transformer. Two months or so before he had arrived in London for his spectacular appearance before the Institution of Electrical Engineers at the end of January 1892, his patron and the institution's president, William Crookes, had stood up to propose a toast at the close of the annual dinner. The toast was 'Electricity in Relation to Science' and Crookes had a good deal to say about it. Crookes at the beginning of the 1890s was at the height of his career, recognized as one of the country's most eminent electrical researchers. His work on the phosphorescence produced in electrical discharge tubes and on the 'fourth state of matter' was highly regarded. He was certainly a man whose views on the electrical future were to be taken seriously. The vision he laid out was one in which the electrical future was only just beginning. It was going to be a future in which the 'vast storehouse of power' locked up in the ether would be released. It would be a future in which the 'almost infinite range of ethereal vibrations or electrical rays' revealed by Hertz in Germany and Lodge in England would be fully exploited for the benefit of all.[1]

Shortly after Tesla's performances at the Royal Institution, Crookes published a greatly expanded version of his after-dinner toast in the *Fortnightly Review*. 'We know little as yet concerning the mighty agency we call electricity,' he stated baldly at the beginning. Was it matter, was it energy, or was it, as Oliver Lodge expressed it, 'a form, or rather a mode of manifestation of the ether'? Whatever it was, the future was going to be dominated by electricity. Crookes invited his readers to imagine a world in which rays of electrical energy could pierce even a London pea-souper. It all raised 'the bewildering possibility of telegraphy without wires, posts, cables, or any of our present costly appliances'. This was 'no mere dream of a visionary philosopher', he said. Crookes described how 'some years ago I assisted at experiments where messages were transmitted from one part of a house to another without any intervening wire by almost the identical means here described.'[2]

Crookes asked his readers to speculate about what the world would look like, seen through eyes that responded to different frequencies of energy. Our eyes respond to a narrow spectrum, but it was 'not improbable that other sentient beings have organs of sense which do not respond to some or any of the rays to which our eyes are sensitive, but are able to appreciate other vibrations to which we are blind'. Those beings would be living in a different world. For them, glass would be 'among the most opaque of bodies'. Any metal 'would be more or less transparent, and a telegraph wire through the air would look like a long narrow tube drilled through an impervious solid body'. To beings who sensed the world like this, a 'dynamo in active work would resemble a conflagration, whilst a permanent magnet would realise the dream of medieval mystics and become an everlasting lamp with no expenditure of energy or consumption of fuel'.[3]

The point of all this imagining, of course, was to hammer home the message that electrical energy was all around us in the ether. What was needed was a way of getting it out. Crookes was clear that he thought that Tesla was at the cutting edge of solving that problem. He had 'succeeded in passing by induction, through the glass of a lamp, energy sufficient to keep a filament in a state of incandescence without the use of connecting wires'. He reminded his readers of some of the effects Tesla had shown off a few days earlier. The 'ideal way of lighting a room would be by creating in it a powerful, rapidly-alternating electrostatic field, in which a vacuum tube could be moved and put anywhere, and lighted without being metallically connected to anything'. Tesla had demonstrated this on the Royal Institution stage, 'by suspending, some distance apart, two sheets of metal, each connected with one of the terminals of the induction coil'.[4]

1892 also saw the publication of the second edition of Oliver Lodge's *Modern Views of Electricity*, first published in 1889. This was Lodge's manifesto for the ether. The ether by now was a simple fact, he told his readers. The ether was no longer a 'hypothetical medium whose existence is a matter of opinion'. Yes, the existence of the ether could be questioned, but only in the way one might cast metaphysical doubt on the reality of matter – the 'existence of an ether can legitimately be denied in the same terms as the existence of matter can be denied, but only so'.[5] Lodge was happy to admit that the details of his physics of the ether might be wrong, but he was as certain that the ether existed as he was certain of the air he breathed. What remained to be done now was properly understanding the nature of the ether, and the 'next fifty years may witness these tremendous victories in great part won'.[6]

The chapter on 'Recent Progress' added to the second edition certainly gave Tesla his due. Hertz's discovery had turned electricity into 'an imperial science' – the key to unlocking the secrets of physics. In the few years since 1888 researchers had searched for more secure ways of detecting the passage of electric radiation through the ether. It had been shown that vacuum tubes 'glow in the oscillating field near an electric vibrator without being attached to wires or any form of conductor'. It was with these kinds of investigation that

'M. Nikola Tesla has delighted every one by constructing alternating dynamos of extraordinary high frequency, which may be considered as a means of maintaining electric oscillations, and thereby making vacuum tubes and other bodies glow with considerably greater brightness.'[7] Tesla's experiments showed just what could be done with the ether as experimenters developed powerful new apparatus – like Tesla's coil itself – that could exploit and harness the energy inside it.

Lodge himself was in the vanguard of attempts to exploit Hertz's discovery. After all he had been in search of electromagnetic waves himself throughout the 1880s and being pipped to the post by Hertz only made his appetite for further wireless discoveries keener. In 1894, following Hertz's untimely death at the beginning of the year, Lodge gave a lecture in his memory at which he demonstrated his work in wireless telegraphy. He had devised a compact Hertz receiver that he demonstrated to the Royal Society in June that year. Telephone receivers turned out to be good detectors of electromagnetic waves as well, and Lodge in his memoirs recalled wandering around the Liverpool University College buildings 'with telephones to my ears, listening to signals from the college'.[8] At the annual meeting of the British Association for the Advancement of Science in August, Lodge showed that he could transmit messages using Morse code by wireless telegraphy from one room to another.

Crookes had mentioned in his *Fortnightly Review* article how he had witnessed some experiments in wireless telegraphy several years earlier. In fact, the experiments he referred to had taken place in 1879, carried out by the Welsh-American inventor David Edward Hughes. Trained originally as a musician, Hughes had made a name for himself as the inventor of the printing telegraph, and had also invented an early form of microphone. Moving from the United States to London in the 1870s, it was while experimenting with his microphone that he noticed that it responded to sparks from an induction coil without any contact between the instruments. He found that he could reproduce the effect at some distance, even receiving the signal in the road outside his house in Great Portland Street. He duly demonstrated the effect to friends, including Crookes as well as his fellow Welshman William Robert Grove. However, Hughes was persuaded by George Gabriel Stokes that nothing more remarkable than ordinary induction was going on, and he took his research no further.[9]

Another witness to Hughes's experiments was William Henry Preece. In June 1892, a few months after playing host to Tesla during his visit to London, Preece had carried out the latest of a series of experiments on the possibilities of wireless telegraphy in South Wales, near Cardiff. In these experiments, lines of insulated cable were laid down on the shore of the Bristol Channel at Lavernock Point, and other lines on the shores of both Flat Holm and Steep Holm islands, in an attempt to communicate between the islands and the Welsh mainland. Preece had been experimenting along these lines since 1885, partly in an attempt – in his capacity as the Post Office's chief telegraph engineer – to investigate the effects of inductance between telegraph lines. His experiments in South Wales were carried out following the establishment of a Royal Commission to investigate the possibilities of electrical communication between the shore and lighthouses and lightships.[10]

Preece carried out a number of similar experiments over the next year or two, including attempts to signal across the Conwy and Dee estuaries in North Wales, as well as across Loch Ness in the Scottish Highlands. Not only could his Loch Ness apparatus send messages by Morse code, but 'speaking by telephone across a space of one and a quarter mile was found practical, and, in fact, easy; indeed, the sounds were so loud that they were found sufficient to form a call for attention.'[11] A very practical, belt-and-braces engineering man, Preece had little time for abstract theory and had been involved in some heated debates with exponents of James Clerk Maxwell's highly mathematical electromagnetic theories over the years. He was firmly of the view that the hands-on experience of electrical engineers was a far more reliable guide to the mysteries of electricity than were the ethereal abstractions of mathematicians.

Preece's most important discovery in wireless telegraphy took place a few years later, however. In July 1896, a young Italian called Guglielmo Marconi arrived outside Preece's office at the Post Office in London. He was clutching a letter of introduction from the Scottish electrical engineer Alan Archibald Campbell-Swinton, and two large bags. In the bags were 'a number of brass knobs, a large sparking coil, and a small glass tube from each end of which extruded a rod joined to a disc fitted to the tube. The gap between the two discs was filled with metal filings.' With the apparatus arranged into two circuits on separate tables, 'Marconi depressed the key in the sparking coil circuit whereupon an electric bell in the coherer circuit rang. Marconi tapped the coherer tube and the bell stopped ringing.'[12] While nothing on those tables was new, Preece was experienced enough to recognize the sensitivity of Marconi's apparatus. This was something that could provide the basis of a practical, working technology.

Marconi was barely 22 years old when he turned up on Preece's doorstep. He was the son of an Italian aristocrat and his Irish wife Annie Jameson. He had been privately educated and had a good grounding in the latest physics thanks to his acquaintance with the Professor of Physics at the University of Bologna, Augusto Righi. It was Righi who alerted Marconi to Hertz's work and to the possibilities of wireless telegraphy that it promised. Using apparatus similar to that developed by Lodge in England, Marconi was soon sending signals from room to room in his parents' mansion in Bologna, and conducted outdoor experiments at his father's country estate. He succeeded in extending the range of his apparatus by using a raised antenna and by grounding the transmitter and receiver. When his efforts to interest the Italian government in his work came to nothing, he travelled to London in search of further patronage and finance.

Even before his appointment with Preece, Marconi had already submitted an application for a patent for 'improvements in transmitting electrical impulses and signals, and in apparatus therefor'.[13] With Preece's help and the resources of the Post Office, he was soon making further progress. It was not long before he was transmitting signals over longer distances. On 27 July he demonstrated his improved apparatus by transmitting from a room in the Post Office building to the roof of another building several hundred yards away, using a sheet of copper as a parabolic mirror to focus the electromagnetic wave, just as a mirror would focus a beam of light. Before the end of the year he had demonstrated his wireless telegraph to the War Office, transmitting a signal over a distance of twelve miles on Salisbury Plain. Preece was convinced that even though there were 'a great many practical points connected with this system that require to be threshed out in a practical manner before it can be placed on the market', it was already clear that it was going to be 'a great and valuable acquisition'.[14]

Preece's growing confidence in his protégé's invention had been further bolstered by more experiments in South Wales. In May 1897 Marconi and his assistants carried out a number of experiments to transmit signals by wireless telegraphy between Lavernock Point on the Welsh mainland and the island of Flat Holm, just as Preece had done a few years earlier. In fact, the first experiment they conducted on 10 May was a repeat of Preece's efforts. Then, on 10 and 11 May they set out to test Marconi's apparatus, transmitting from

Flat Holm using an induction coil that generated an impressive 20-inch spark as their transmitter. It almost didn't work: 'On the 11th and 12th his experiments were unsatisfactory – worse, they were failures – and the fate of the new system trembled in the balance. An inspiration saved it. On the 13th the receiving apparatus was carried down to the beach at the foot of the cliff, and connected by another 20 yards of wire to the pole above, thus making a height of 50 yards in all. Result, magic! The instruments, which for two days failed to record anything intelligible, now rang out the signals clear and unmistakable, and all by the addition of a few yards of wire! Thus often, as Carlyle says, do mighty events turn on a straw.'[15]

Those experiments proved to be decisive. They were followed a few days later by further successful attempts at signalling between Lavernock and Brean Downs on the other side of the channel. By now, accounts of the new Marconi system of wireless telegraphy were beginning to appear in the press. The *Daily News* reported in June that according to Preece: 'Marconi messages have been sent between Penarth and Bream Down, near Weston-super-Mare, across the Bristol Channel, a distance of nearly nine miles.'[16] A few days later, the *Graphic* was reporting that 'we can be sure, that the investigations of the last few years have opened up a region which will yield far more remarkable phenomena than any now within the knowledge of man.'[17] A year later Marconi managed a particularly clever publicity stunt when he fitted up a boat with wireless telegraph apparatus and provided running commentary on the Kingstown Regatta for the Dublin *Daily Express*. He repeated the trick a year later, covering the America's Cup for the *New York Herald*. The coup gained him and his wireless telegraph their first toehold in the American market.

In the meantime, others were making their own efforts to reproduce Marconi's experiments. One of the participants in the

Flat Holm trials had been the German professor Adolf Slaby, who had been deeply impressed by seeing how the signals 'came to us dancing on that unknown and mysterious agent the ether!'.[18] Back in Potsdam, Slaby carried out his own experiments, including some carried out before the Kaiser himself. He was soon sending signals as far as 21 kilometres, using antennae hoisted 300 metres into the air by balloon.[19] In March 1899 Marconi succeeded in sending wireless messages across the English Channel to France. The *Daily News*'s correspondent waxed lyrical about 'how those messages were flashed or winged by Signor Marconi and his wonderful system of wireless telegraphy'. By now it almost seemed routine, as 'the thing was done with as much accuracy and clearness as if the words had come by wire instead of being dispatched through the empty air.' The 'magic of it all naturally impresses one who for the first time sees the little there is to see', but for Marconi it was all 'a matter of course'.[20]

Tesla, however, was not at all impressed. He simply could not see that there was anything novel about what Marconi had done, he rather tetchily told the *Daily News* correspondent. 'While Signor Marconi's experiments are most interesting, they are not novel,' he complained. The apparatus that Marconi used was just the same as the arrangement described by Oliver Lodge in 1893, and which Tesla himself had also demonstrated in a lecture that year. It also suffered from the defect that any signals using such apparatus could be destroyed by the 'introduction of an induction coil anywhere between the terminals, or anywhere in the neighbourhood of one of them'. As a result, Marconi 'can do nothing commercially'. Tesla had in fact abandoned this entire approach to wireless telegraphy as being worthless. Improving the original Lodge apparatus had proven impossible, so: 'I abandoned it, and have since devoted myself to perfecting another apparatus, on which I have worked for a long time.' 'I do not wish to detract from Signor Marconi's success,' he concluded, 'but I do not see why there need be much stir about it.'[21]

Like Marconi, Tesla had been inspired by Hertz's researches; the experiments that led to the oscillating transformer had been the outcome of that inspiration. But Tesla also thought that these experiments were the key to a far grander and more ambitious transformation of the future than anyone else had yet imagined. By the middle of the 1890s Tesla had carefully cultivated for himself an image as a dreamer of electric dreams. He was the 'inventive electrical poet, who takes hold of an idea, and with philosophy as a fulcrum tries to unlock the universe'.[22] In the wake of his work on high-frequency oscillations in the early years of the decade, he was dreaming about sending not just signals, but power everywhere, and through the earth rather than the ether. It would be possible 'by means of powerful machines' to disturb the 'electrostatic condition of the earth, and thus transmit intelligible signals and perhaps power'. 'In fact,' he asked, 'what is there against the carrying out of such a scheme?'[23]

CHAPTER 13

The Pursuit of Power

Clearly, though Tesla might not have thought much of Marconi's efforts at wireless telegraphy, he had great ambitions for his own wireless world. Those ambitions were clear in the immediate aftermath of his triumphant lecturing performances in New York in 1891 and in London and Paris at the beginning of 1892. He had spoken at those lectures about the possibilities of transmitting energy wirelessly and had been further encouraged by the enthusiastic reception his speculations had received. Not only had the popular and electrical press been fulsome in their plaudits, but eminent men of science like Crookes, Lodge, Preece, and even Lord Rayleigh had expressed their admiration for his ideas. Tesla had been particularly struck by Rayleigh's praise. The former Cavendish Professor of Physics and Professor of Natural Philosophy at the Royal Institution had delivered a speech of thanks after his lecture at that venerable establishment, and had lauded Tesla's impressive capacity for scientific imagination. His words convinced Tesla that he had something unique to offer and that he should look for 'some big idea' as the focus for his genius.[1]

A year or so later, in lectures in Philadelphia and St Louis, Tesla set out the progress of his vision. Many of the spectacular experiments he demonstrated at these lectures would have been familiar to those who had read about his performances the previous year. This time, however, he expressed himself in far more apocalyptic terms. 'The day when we shall know exactly what "electricity" is,' he told his audience, 'will chronicle an event probably greater, more important than any other recorded in the history of the human race.' The time was coming when 'the comfort, the very existence perhaps, of man will depend upon that wonderful agent.' The future would depend on energy, harnessed in places like Niagara – and 'how will they transmit this energy if not by electricity?' Tesla acknowledged that 'this view is not that of a practical engineer.' But it was not just fantasy either: 'neither is it that of an illusionist, for it is certain, that power transmission, which at present is merely a stimulus to enterprise, will some day be a dire necessity.'[2]

Tesla was both more ambitious and more confident than he had been in London and Paris about the prospects for his wireless vision of the future. He now talked explicitly about 'the transmission of intelligible signals or perhaps even power to any distance without the use of wires'. It might still be the case that 'the great majority of scientific men will not believe that such results can be practically and immediately realized', but Tesla had no doubts. His confidence was such that 'I no longer look upon this plan of energy or intelligence transmission as a mere theoretical possibility, but as a serious problem in electrical engineering, which must be carried out one day.' It was 'the natural outcome of the most recent results of electrical investigations'.[3] The St Louis lecture in particular was a huge success, with 4,000 people in the audience. When it was all over, 'so great was the desire of the public to see Mr. Tesla closer, an informal reception was held in the lobby, when several hundred of the leading citizens seized the opportunity and Mr. Tesla's hand in a very vigorous fashion.'[4]

Thomas Commerford Martin's account of Tesla's researches in the *Century Magazine* a couple of years later reinforced the message that he was on the verge of a wireless breakthrough. Martin was clear to his readers that the 'vast new electrical domain which the thought and invention of our age has subdued' was on the verge of another transformation. It was a transformation with an uncertain outcome, since 'one can never be sure into what part of the social or industrial system it is next to thrust a revolutionary force.' As far as electricity was concerned, the 'fanciful dreams of yesterday are the magnificent triumphs of tomorrow, and its advance towards domination in the twentieth century is as irresistible as that of steam in the nineteenth'.[5] This was all part of a calculated campaign to make sure that, as far as the American public was concerned, Tesla was recognized as the key mover and maker of this new electrical future. Martin, though they were soon to fall out, was in the mid-1890s one of Tesla's most enthusiastic backers.

The article was packed with examples of Tesla's ingenuity and his mastery of wireless technology. Tesla's oscillator would revolutionize the provision of electric light by working directly through the ether. His apparatus 'gets the ether medium into such a state of excitement in which it seems to be capable of almost anything'. No one who saw the experiments could 'fail to be impressed with the actuality of a medium, call it ether or what you will, which in spite of its wonderful tenuity is as capable of transmitting energy as though it were air or water'. Even more remarkable, Martin argued, was the mastery that his instruments allowed Tesla to demonstrate over this flow of energy through the ether. All this, 'aside from their deep scientific import and their wondrous fascination as a spectacle', pointed to 'many new realizations making for the higher welfare of the human race'. The 'transmission of power and intelligence' was one prospect; the 'modification of climate' another.[6]

By now though, Tesla was sure that sending energy through the ether alone was not the solution he was looking for. He was thinking instead of the possibilities of transmitting electricity through the earth, with the atmosphere acting simply to complete the circuit. With an oscillator that had been modified in order to maximize the ground current, 'if he has not yet determined the earth's electrical "charge" or "capacity",' wrote Martin, 'he has obtained striking effects which conclusively demonstrate that he has succeeded in disturbing it.' He was pumping energy into the earth just like 'a pump forcing air into an elastic football'. If the oscillator's frequency resonated with the natural frequency of the earth, then the 'purple streamers of electricity thus elicited from the earth and pouring out into the ambient air are marvellous'.[7]

Tesla carried on worrying away at his grand project for the next several years. The idea was in principle a simple one. If electrical energy could be pumped into the earth at the right frequency it could be transmitted to any distance and returned through the atmosphere, completing a kind of wireless circuit. During his experiments to improve his apparatus, Tesla recognized that gases became better conductors of his high-frequency transmissions when they were at lower pressures. He noted that with high enough voltages and frequencies, and low enough pressures, it seemed possible that huge quantities of energy could be transmitted over indefinite distances. The way to do this in practice was to ensure that the return current was confined to the upper atmosphere where air pressure was lowest. Tesla filed a patent application for this new system in September 1897. He described his apparatus for generating electricity and how one 'terminal of this generating apparatus is connected to earth, and the other terminal is maintained at an elevation, where the rarefied atmosphere is capable of conducting freely the particular kind of current used'.[8]

Within a year of filing his patent application for wireless electrical energy transmission, Tesla had gone west. In May 1899 he abandoned his New York base and moved his operations almost 2,000 miles away to Colorado Springs. The choice of location was deliberate. In order to perfect his apparatus Tesla felt that he needed to locate his laboratory at a high altitude where he could take advantage of the more rarefied atmosphere he needed for his experiments. Tesla made it clear to the *Denver Rocky Mountain News* that he was not on holiday. He 'intended to make extensive experiments in this altitude, or rather, at the top of the peak'. He 'had come here for work and not for pleasure nor for his health'.[9] A few days later the newspaper reported that Tesla had leased a piece of ground behind the State School for the Deaf and Blind, 'where he will construct a station for the purpose of carrying on his experiments'.[10]

Just as Tesla was setting up shop in Colorado Springs, an article in *Pearson's Magazine* offered its readers a tantalizing glimpse of the future that Tesla would be trying to turn into reality there. No one, declared Chauncey Montgomery McGovern, 'can escape a feeling of giddiness when permitted to pass into this miracle-factory and contemplate for a moment the amazing feats which this young man can accomplish by the mere turning of a hand'. Readers were introduced to a 'large well-lighted room, with mountains of curious looking machinery on all sides', and a 'tall, thin young man' who 'by merely snapping his fingers creates instantaneously a ball of leaping red flame, and holds it calmly in his hands'. This was Tesla, the maker of the future, whose plan to 'harness the rays of the sun to do man's bidding is probably the boldest engineering feat that he or anyone else has ever contemplated'.[11]

Tesla's future was one in which steam directly generated by the sun's heat using lenses and mirrors would be used to generate unprecedented electrical energy that could then be wirelessly transmitted to anywhere there was an appropriate receiver to receive it. This was the fantasy that the Colorado Springs laboratory was meant to turn into reality. The aim was to build power stations topped by large towers, above which would float balloons connected to the towers by cables. The electricity generated below that would

Tesla demonstrating wireless illumination. (*Pearson's Magazine*, 1899)

be 'thus set free will be carried on by the atmosphere to any indefinite distance'. There would also be a network of receiving stations, with towers and balloons 'which will be equipped with the apparatus necessary to absorb the free electricity in the atmosphere and send it to the receiving station below', where it could be distributed as required. McGovern had not quite understood that Tesla's plan was to transmit electricity through the earth rather than the heavens, but the image he evoked of networks of transmitting towers and balloons sending energy back and forth over great distances was exactly that described in Tesla's 1897 patent.

Tesla's proposed arrangement of balloons for the wireless
transmission of electrical power. (*Pearson's Magazine*, 1899)

By the end of June, newspapers were reporting that Tesla's laboratory was ready for experiments to begin. The *Colorado Springs
Evening Telegraph* announced that 'the apparatus necessary to successfully conducting his experiments has been received by Nickola

[*sic*] Tesla, and his experimental station near the Printers' Home is taking on an air of activity.' The place was 'filled with dynamos, electric wires, switches, generators, motors, and almost every conceivable invention known to electricians, and through this mass of intricate and dangerous mechanism Mr. Tesla walks as fearlessly as if on the streets of the city'. The wooden building was 60 feet wide and 70 feet long and was topped by a retractable 142-foot-long mast with a copper ball at its apex. Inside was a huge magnifying transmitter consisting of a primary coil made of two turns of thick cable and a secondary coil inside it consisting of a hundred turns of finer cable. The secondary coil was grounded and could be attached either to a terminal inside the building or to the copper ball on the mast above.[12]

Tesla's first task was to attempt measurements of the earth's electrical potential, free from the already considerable electrical clutter of New York. Accordingly, Tesla grounded one end of the primary coil while attaching the other end to 'an elevated terminal of adjustable capacity' (presumably the copper ball on top of the mast). Any changes in the earth's potential would cause surges of current in the primary, which would in turn cause magnified surges of current in the secondary coil that would be recorded by his detector. Tesla was delighted with the outcome: the 'earth was found to be, literally, alive with electrical vibrations'.[13] Tesla was also struck by the observation that the numerous thunderstorms in the vicinity of the station sometimes seemed to have a greater effect on his apparatus when they were further away. Experimenting further, Tesla arrived at a startling conclusion. What he was seeing was evidence of stationary electric waves in the earth.

This evidence of stationary waves seemed to Tesla to offer the prospect of an entirely new system of wireless transmission. If the earth behaved 'like a conductor of limited dimensions' then it was

The Tesla Collection

The Colorado Springs experimental station.

possible 'to impress upon the entire globe the faint modulations of the human voice, far more still, to transmit power, in unlimited amounts, to any terrestrial distance, and without any loss'.[14] Tesla spent the next several months at Colorado Springs further refining his apparatus and dreaming about the possibilities for energy transmission that they offered. He was convinced that he was on the verge of discoveries without precedent – and certainly that he had a system far more powerful and superior in every way to Marconi's wireless telegraphy. Now he needed money to turn his dreams into a reality. Back in New York by the turn of the century, Tesla was soon busy drafting new patent applications, and looking for new financial backers.

He was also working on a manifesto. For much of the past decade he had been working hard at selling his vision in lectures and interviews with the electrical and popular press. Now that this vision was complete – in his eyes at least – he wanted an opportunity to set it out in detail. Tesla was convinced that he held the future in his hands. His new system of energy production and distribution was not only going to make him a fortune and establish him as the century's greatest inventor, it was going to transform the future organization of human society. This time Tesla wanted to appeal directly to the broader public rather than to fellow electricians. He wanted the tide of public opinion to add weight to the economic case he would need to present to potential investors. The *Century Magazine* must have seemed the obvious place to publish. One of his few close friends, the journalist Robert Underwood Johnson, was an associate editor, and the magazine had already published a substantial and highly complimentary account of his work, written by another friend, Thomas Commerford Martin, just a few years previously.

Tesla called his essay 'The Problem of Increasing Human Energy'. He introduced his thoughts by musing on 'that inconceivably complex movement which, in its entirety, we designate as human life'. Movement was the defining feature of the universe, not just through space, but through time: 'its destination is hidden in the unfathomable depths of the future.' Tesla meant this literally, as well as metaphorically. The universe existed through movement and life needed movement to thrive. He invited his readers to think about it like an equation; to think of man as a 'mass urged on by force'. Maintaining the forward movement of this mass needed energy, and so 'the great problem of science is, and always will be, to increase the energy thus defined.'[15] That might be done by increasing the mass of humanity itself, or by reducing the forces that opposed humanity's onward momentum. Or the problem might be solved by increasing the sources of human energy itself. It was an argument that appealed explicitly to the ideal of progress that dominated late nineteenth-century intellectual culture.[16]

Human mass might be preserved and increased in all sorts of ways: 'by careful attention to health, by substantial food, by moderation, by regularity of habits, by the promotion of marriage, by conscientious attention to the children, and, generally stated, by the observance of all the many precepts and laws of religion and hygiene.' To maximize the increase of human energy it was essential that successive generations improve upon the previous one. Only if each generation were 'further advanced' in culture and intellect would they 'add very substantially to the sum total of human energy'. Similarly, reducing the forces that opposed the increase of energy meant combating 'ignorance, stupidity, and imbecility', as well as 'visionariness, insanity, self-destructive tendency, religious fanaticism, and the like'.[17] Tesla advocated technologies to do away with war, like flying machines, and automata that would make war 'a mere contest of machines without men and without loss of life'.[18]

The real solution to the problem of increasing human energy, though, was finding better ways of harnessing the energy of the sun and other natural sources of energy. The sun was the 'spring that drives all', so at the most fundamental level, to 'increase the force accelerating human movement means to turn to the uses of man more of the sun's energy'. Electricity was the key to doing this, both in terms of the most efficient ways of generating energy and the most efficient means of transmitting energy to wherever it was needed. It was vital to 'recognize the transmission of electrical energy to any distance through the media as by far the best solution of the great problem of harnessing the sun's energy for the uses of man'. It was Tesla's own work on wireless transmission at Colorado Springs that made that solution viable. Tesla could 'conceive of no technical advance which would tend to unite the various elements of humanity more effectively than this one, or of one which would more add to and more economize human energy'.[19]

Tesla's vision of the future was received by some commentators in the electrical press, at least, with a mixture of derision and hilarity. The article 'does not contain one exact measurement or statement of an attained result', and was 'mainly of interest on account of the pretty and somewhat startling photographs', said one.[20] The *Electrician* joked that journalists were owed an apology from those who in the past might have accused them of 'imaginative embroidery', since 'this contribution equals in wild speculation anything hitherto attributed to the eminent inventor'.[21] Contributors to the *Electrical Review* or the *Electrician* were not the audience that Tesla had in mind for these latest arguments, however. He clearly felt that he did not need the approval or support of electricians any more. Now he wanted to be more than just an inventor of technology – he wanted to invent the future. Tesla's experiments at Colorado Springs had convinced him that there was more to electricity than mere technology. Now he needed to sell that conviction: to the public and to financiers anxious to offer that public what it wanted.

CHAPTER 14

Other Worlds

Tesla's reinvention of himself as a new kind of inventor at the end of the 1890s was the culmination of a long process. At the beginning of the decade he had been largely invisible beyond the relatively small circle of professional electrical engineers. He was one of a number of electrical inventors struggling for recognition by his peers and competing for the attention of investors. It was his public performances of the early 1890s that changed that. Those performances, and the press coverage they received, made Tesla into a public figure. Increasingly in the years that followed, his activities and pronouncements became topics of attention. Newspapers reported his views on the latest scientific developments. Magazine editors commissioned articles about him and his inventions. Tesla himself actively courted this kind of attention, and clearly enjoyed the trappings of fame. He dined at the famous Delmonico's restaurant, rubbing shoulders with New York's fashionable elite. In 1898 he would move in to rooms in the luxurious Waldorf Astoria hotel. He befriended celebrities like Mark Twain. This was how he thought the inventor of the future should live. Tesla was learning that projecting a particular image of himself would be a key element in the process of making that future a reality.

Reporting on Tesla's performances in New York, London and Paris in 1892, the *New York Sun* remarked that 'even in its present state, the Tesla apparatus must rank with Franklin's kite ... with Bell's disks ... and with Hertz's resonator.' His experiments showed 'that every common flame is but a brilliant manifestation

Mark Twain playing with an electrical ball
of light. (*Century Magazine*, 1895)

of electro-magnetic energy'. Tesla himself was 'the Columbus of a new continent for inventive activity'.[1] A year later the *New York Herald* informed its readers that 'Nikola Tesla has been called by scientific men, who do not award praise freely, or indiscriminately, "the greatest living electrician".' His lecture at the St Louis electrical convention, in the 'Grand Music Entertainment Hall, was listened to by a larger audience than had ever been gathered together before in the United States on an occasion of this kind.'[2] He was hailed as the genius of Niagara and the power behind the Columbian Exposition. Tesla the electrician was being turned into an American icon.

This was not an accidental process. Tesla himself and some of his promoters – his friend Thomas Commerford Martin in particular – were fully complicit in efforts to keep him in the public eye. They were keen to make sure that he was seen as a man of science rather than a mere maker of machines. Tesla was a visionary with the capacity to gaze more deeply than others into the mysteries of nature. Much was made of his exoticism (though his exotic origins were not always accurately located: one reporter described him as Hungarian, for example, and another thought he was Italian). In 1894 Martin published an edition of *The Inventions, Researches and Writings of Nikola Tesla* as part of a campaign to bring his protégé further publicity. The *New York Times* review hailed Tesla as someone who possessed 'a genius for electrical work "with all the play and push of an original mind".'[3] Martin's essay on Tesla in the *Century Magazine* a few months later was a further effort to mould his public image.[4]

Disaster struck for Tesla and his ambitions when in the early morning of 13 March 1895 a fire broke out in his laboratory, destroying it completely. He lost thousands of dollars' worth of equipment and, even worse, his notes on his experiments. This too was enthusiastically reported upon by the press, making headlines across North

America and Europe. Tesla was 'utterly disheartened and broken in spirit,' reported the *New York Herald* – 'one of the world's greatest electricians' had taken to his bed in nervous prostration. The 'web of a thousand wires which at his bidding thrilled with life had been tangled by fire into a twisted skein'.[5] Tesla's loss, according to the *New York Sun*, was 'a misfortune to the whole world'. It told its readers 'that the men living at this time who are more important to the human race than this young gentleman, can be counted on the fingers of one hand; perhaps on the thumb of one hand'.[6] London's *Electrical Review* reported that the 'damage is estimated at $100,000, but the greatest loss is the physical collapse of the inventor from the shock caused by the disaster'.[7] Typically, Tesla turned to electricity to deal with the depression that followed the fire. He took to administering regular doses of shocks to himself. Electricity could restore the flagging nerve force, depleted by the strain that the shock and frantic overwork had imposed on his system.

Deliberately or not, the image of the inventor being conveyed in these and other, similar accounts of Tesla throughout the 1890s was a far cry from the robust public face of an Edison, for example. There was something not quite worldly about Tesla. Even while newspapers were filling their pages with accounts of his inventive exploits, they were telling their readers that Tesla 'doesn't like to see his name in print'. The inventor wanted 'more than anything else, to be left alone'.[8] It was a public image that played with the idea of the sage in solitude rather than that of the hard-headed practical man. He was 'the modern Benjamin Franklin', who had 'sent his electrical kite to the sky for studying the secret of light'. The writer Francis Leon Chrisman thought that, like Cassius, 'he has a "lean and hungry look", and "thinks too much".' Tesla was 'dreaming daily of new worlds to conquer'.[9]

Whatever his might say about wishing for solitude, Tesla was clearly happy to provide newspapers with material. In turn, the newspapers were happy to lap it up. Tesla the otherworldly man of science evidently made for good copy. His movements were news – his relocation to Colorado Springs in 1899 was a topic of excited speculation. His reputation (to the American public at any rate) as the greatest living electrician allowed him to repeat his view that there was 'nothing really novel, you understand, about what Marconi is doing'. His experiments were 'interesting, that is all'.[10] Tesla's self-made public image could sometimes backfire on him, nevertheless. 'America's Own and Only Non-Inventing Inventor, the Scientist of the Delmonico Café and Waldorf-Astoria Palm Garden, has been at it again,' quipped one commentator. As far as that particular commentator was concerned, there was in the end only one real difference between Marconi and Tesla – 'the actual results of the methods of the two inventors show only this slight difference: Marconi telegraphs through space and Tesla talks through space.'[11]

Wilhelm Röntgen's discovery of X-rays at the end of 1895 provided Tesla with plenty of opportunities to remind the public of his mastery of electromagnetic mysteries. As soon as he heard of Röntgen's discovery, shortly after it was announced, Tesla was convinced that he had only just missed making the discovery for himself. A few months earlier he had been trying to take photographs using a glowing Crookes tube as the source of illumination and had been disappointed to see that the photographic plate had been spoilt – maybe that fogging had been caused by Röntgen's mysterious rays. He was soon feverishly experimenting and in March the *New York Times* could announce that the results of Tesla's efforts would soon appear in the *Electrical Review*. He had succeeded in producing X-ray photographs of unprecedented clarity, and was 'happy to have contributed to the development of the great art' that Röntgen had created.[12] The *New York Herald* had already reminded their readers that 'all the discoveries made through the popularization of the Roentgen experiments were made possible by a remarkable invention of Mr. Tesla – his converter.'[13]

Throughout the remainder of 1898, Tesla's investigations of X-rays and his speculations both about their nature and their future uses provided ample material for the newspapers. As well as his X-ray research, throughout 1898 Tesla had been working away at another ambitious project – to control torpedoes by wireless. He had developed a system through which a torpedo boat could be manoeuvred up to enemy vessels and exploded by remote control. Throughout the 1890s, the European powers had been competing with each other to build new, more heavily armed and more heavily protected warships. Tesla was confident that in such a climate there would be a market for weaponry that could challenge the supremacy of these steel-clad behemoths without risk. The Spanish American war that broke out in April that year also focused minds on the United States' capacity for naval warfare. Tesla applied for a patent for his invention in July and received it in November.

With patent secured, Tesla embarked on a concerted publicity campaign. The *New York Journal* carried news that the 'great magician of electricity' had invented a submarine boat that would 'carry no lives to risk, but can be directed from on shore or from the decks of a war ship'.[14] Tesla announced that with his weapon 'war would be abolished.' Navies would be useless in the face of a weapon that could be deployed at a distance. Armadas could be wiped out before they set out for sea – it would 'work a revolution of the politics of the whole world'.[15] The invention would 'make war so terrible, as well as expensive, as to make it prohibitory, and thus to assure peace between all the nations'.[16] Tesla promised that in a couple of years' time his invention would by on show at the 1900 Paris Exhibition,

A demonstration of Tesla's torpedo boats controlled by
wireless. (*New York Herald*, 8 November 1898)

with the torpedoes being controlled by him wirelessly across the
Atlantic from his laboratory in New York.[17]

This relentless campaign of self-promotion was not without its
cost. There were murmurs that even by Tesla's standards of sensationalism this was a step too far. The day after the *New York Herald*
published its paean of praise, it published an interview with Cyrus
Fogg Brackett, Professor of Physics at Princeton University, that
poured scorn on Tesla's pretensions. As far as he was concerned,
'what is new about it is useless, while that which is useful had all
been discovered long before Tesla made this startling announcement.'[18] Brackett's criticism was not the only one. The *Scientific
American* noted that they were 'not alone in our expressions of regret
that any one of Mr. Tesla's undoubted ability should indulge in
such obvious and questionable self-advertisement'. Most seriously,
the business brought about a permanent rupture between Tesla

and Thomas Commerford Martin. In an editorial in the *Electrical Engineer* an exasperated Martin complained that 'Mr. Tesla fools himself, if he fools anybody, when he launches forth into the dazzling theories and speculations associated with his name.'[19]

Tesla was furious at Martin's betrayal and responded in kind. Past patronage was laid entirely aside in response to the 'serious injury' Martin had inflicted. In the past 'both as Christian and philosopher I have always forgiven you and only pitied you for your errors.' But this time 'your offence is graver than the previous ones, for you have dared to cast a shadow on my honor.' One of the reasons Tesla was so angry was that he really did believe his own hyperbole. The possibilities of his remote-controlled weaponry and similar devices would play a central role in his formulation of the 'problem of increasing human energy' a couple of years later, appearing under the heading 'The Art of Telautomatics'. The key to eliminating the destructive potential of warfare was to produce a future in which 'machine must fight machine' rather than humans fighting each other. It would need 'a machine capable of acting as though it were part of a human being ... as though it had intelligence, experience, reason, judgement, a mind!'. His remote-control weapons were a step in that direction because they embodied the minds of their operators and allowed them to act at a distance.[20]

In an interview with the *New York Herald* back in the summer of 1896, Tesla had speculated about the possibilities of seeing at a distance. 'This problem of transmitting sight by wires or otherwise to any distance is as difficult as it is fascinating,' he told the newspaper. It was a problem that had exercised his inventive imagination too, he confessed, though 'while in these other lines I have considerably realized, as regards this problem of transmitting sight, I am still very far from positive demonstration by experiment.'[21] As Tesla himself noted, there had been much talk about telectroscopes, as devices for seeing at a distance were commonly called, in the aftermath of Bell's invention of the telephone – if the one was possible, why not the other? A few months after Tesla's interview, the Polish inventor Jan Szczepanik took out a British patent for such a device and there was a great deal of speculation surrounding his invention. In 1898, Tesla's friend Mark Twain would write a short story for the *Century Magazine* exploring the possibilities of Szczepanik's invention. As with Tesla and his wireless weaponry, Szczepanik promised to put his invention on show at the 1900 Paris Exhibition. Neither invention would appear there.

Throughout these sorts of discussions about the potentials and possibilities of future technologies, the line between fictional and factual speculation was an extremely narrow one. Telling tall tales about the futures that new technologies might lead to was part of the business of invention. Tesla clearly understood this very well and devoted a great deal of attention throughout the 1890s to giving his public tantalizing glimpses of the future world his inventions might bring about. Edison was another master of such speculation, and sensationalist newspapers like the *New York Herald* thrived on bringing hints of those possible future worlds to their readers. Clearly, there were some commentators, like Thomas Commerford Martin, who thought that too much speculation lowered the dignity of the inventor. Others, like Mark Twain in his short story about Szczepanik and his telectroscope, were happy to play artfully with the boundary between fact and fiction. Tesla's speculations were playing with that boundary too.

It would therefore have come as little surprise to Tesla's followers when he began discussing the possibility of employing his inventions to send communications to Mars. The Red Planet and its inhabitants were topics of excited speculation throughout the second half of the decade. In 1895 the American astronomer Percival

Lowell had published *Mars*, containing maps and vivid descriptions of the 'canals' he had seen on the planet's surface during observations at his Flagstaff Observatory. Lowell was sure that the features he saw criss-crossing the Martian surface were artificial in origin and that they were signs of the last gasps of an ancient civilization. Similar claims had been made by Camille Flammarion and Giovanni Schiaparelli. The book, and Lowell's inferences about life on Mars and its nature, caused a sensation. If there was not just life but civilization on Mars, could they observe human civilization in the same way that Lowell had observed theirs? Was travel to Mars possible, or some other form of communication?

Lowell's speculations were factual; those of the budding English author H.G. Wells were explicit fiction. In 1898 Wells published his *The War of the Worlds* in novel form. It had appeared originally as a serial in *Pearson's Magazine* the previous year, and pirated in the *New York Evening Journal*, which changed the setting from London to New York. In the context of the excited speculation about Martian civilization and its possibilities that had been sparked by Lowell's discoveries only a few years previously, Mars was the obvious choice as the invading aliens' point of origin. Speculations about the age of Martian civilization and the way in which Lowell, like others, characterized the Martians themselves as a dying race in an increasingly hostile environment, offered a logic for invasion. Stories like *The War of the Worlds* worked for readers because they appealed to what they had already been told about Martian culture, just as they contributed to growing speculation about what Mars and the Martians were really like.

In view of all this, it is no surprise that the possibilities of communication with Mars played such a prominent role in discussions of Tesla's plans for wireless communication. As early as 1896 Tesla was talking up the possibilities of communicating with Martians in an interview with the *New York Sun*. 'I have had this scheme under consideration for five or six years,' he told the newspaper, 'and I am becoming more convinced every day that it is based upon scientific principles and is thoroughly practicable.'[22] On his way from New York to Colorado Springs in May 1899, Tesla stopped off in Chicago to lecture to the city's Commercial Club. Asked by the *Chicago Times-Herald* about the prospects for Martian communication, he said: 'I have apparatus that can accomplish it beyond any question. If I should wish to send a signal to that planet I could be perfectly certain that the electrical effects would be thrown exactly where I desire to have them and that the exact signals I desire to make would be made.'[23]

'Mr. Tesla has not yet attempted to signal to any point so far away as Mars,' the interviewer told his readers. By the beginning of 1901, however, speculation was mounting that Tesla had received signals from Mars during his Colorado Springs experiments. The *New York Journal and Advertiser* had him writing that 'I have observed electrical actions which have appeared inexplicable, faint and uncertain though they were, and they have given me a deep conviction and foreknowledge that ere long all human beings on this globe, as one, will turn the eyes on the firmament above with feelings of love and reverence, thrilled by glad news.'[24] In less apocalyptic terms, the *New York Sun* a few days later described how 'Tesla has been able to note a novel manifestation of energy, which he knows is not of solar or terrestrial origin, and, being neither, he concludes that it must emanate from one of the planets.'[25]

Two months later, Mars and Tesla were on the front pages again. In March, the boy's magazine *New Golden Hours* started serializing a novel titled *To Mars with Tesla*, written by the pulp fiction author Weldon J. Cobb. Edison had already visited the planet, in a novel called *Edison's Conquest of Mars* written in 1898 by Garrett P. Serviss

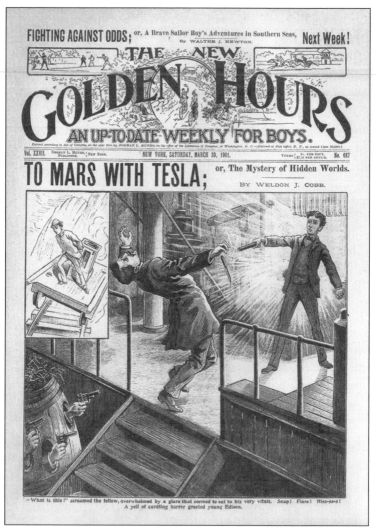

Tesla was the hero of this boys' adventure story:
'To Mars With Tesla'. (*New Golden Hours*, 30 March 1901)

as a sequel to Wells's *The War of the Worlds*. Cobb, a prolific author, had presumably recognized the opportunity for a Martian visit by Tesla in the light of headlines about his communications with the planet. Indeed, the adventure started with Tesla's attempts to build a device to send messages to Mars and attempts to thwart him by a series of villains in possession of a variety of flying machines. Edison was clearly a willing participant in his fictional representation. It is less clear to what extent Tesla knew about or approved of his own foray into scientific romance, though it seems unlikely that he did not know about it.

Tesla by the beginning of the new century was the American representative of sensational science. He had successfully positioned himself as the scientific inventor to whom journalists could turn to for useful soundbites on the latest discoveries. He offered the readers of sensation-seeking newspapers and middlebrow magazines like the *Century* or *Pearson's* a compelling vision of the future his own inventions would bring about. Tesla's future was going to be one where America's cities would be powered by Niagara, its energies transmitted wirelessly through earth and air. It was going to be a future of wireless communication between continents – and between worlds. It was going to be a future where war was made using automata rather than people. But Tesla also understood very well that making this future real would need more than sensationalism. The sensationalism that underpinned his public image at the beginning of the twentieth century was simply a means to an end. By selling sensational science to the public he hoped to be able to sell himself to investors too.

CHAPTER 15

Wardenclyffe

Back in New York from Colorado Springs at the beginning of 1900 – and busily penning his manifesto for the *Century Magazine* – Tesla was also thinking about how to turn his vision of the future into reality. He knew that he needed patents and he knew that he needed powerful financial backers. Brimming with confidence, he told the press that his experiments had 'been most successful, and I am now convinced that I shall be able to communicate by means of wireless telegraphy not only with Paris during the Exhibition, but in a very short time with every city in the world'. He was back in New York, he made clear, 'to pursue a series of experiments with a view to solving the greater problem of transmitting power without wires'.[1] Those Colorado Springs thunderstorms and their effect on his apparatus had convinced him that he really could establish stationary electromagnetic waves inside the earth. He simply needed to make his instruments even bigger. As he worked on the patents that would be needed to protect his system and to establish his rights to potential investors, Tesla bombarded the press with promises about the electrical future he could offer.

Over the next few months Tesla filed patents for the 'Art of Transmitting Electrical Energy through the Natural Mediums'; a 'Method of Signalling'; a 'System of Signalling'; and a 'Method of Insulating Electrical Conductors'.[2] In March, the electrical press reported that he had been awarded his patent for wireless transmission through a rarefied atmosphere that he had filed before setting off for Colorado Springs.[3] The publication of his long essay on 'The

Problem of Increasing Human Energy' in the *Century Magazine* in June generated another flurry of publicity, not all of it favourable, particularly in the scientific press. In August, the *New York Herald* reported that a New York physician, Dr Craft C. Carroll, had used one of Tesla's oscillators to cure an otherwise incurable case of tuberculosis.[4] Later in the month, the *New York World* reported that Tesla himself had instigated the experiment.[5] By the end of the year the press was reporting that thanks to Tesla's inventions, 'Men of the Future May Become as Gods'.[6]

News of Tesla's return to New York and his aspirations for the future were reported on the other side of the Atlantic as well.[7] The Guernsey *Star* listed Tesla's work on 'a method of transmitting power electrically through the rare upper air', along with 'the long-looked-for discovery of the means of transmitting vision over electric wires' by 'a young Pole named Szczepanik' among the great scientific discoveries of the previous year.[8] The *Pall Mall Gazette* reported Tesla's article in the *Century Magazine* in particularly glowing terms. 'Lord Lytton in "The Coming Race" and Mr. H.G. Wells in "The War of the Worlds" and "When the Sleeper Wakes" have both drawn pictures of what the world will be like when the scientific inventor really lets himself go,' it told readers. But if 'Mr. Tesla succeeds in making half his discoveries available for daily use, we shall have everything at our command that the Vrilya had, and shall have gone a long way towards acquiring the amazing forces of the Martians'.[9]

Now it was time for Tesla to start looking for that vital financial backing. George Westinghouse was one of the first potential investors he approached for help in turning the experimental apparatus he had developed at Colorado Springs into a commercial proposition. As a former patron and partner, Westinghouse was an obvious choice in this respect. He had already invested in Tesla's dreams before, after all, and Tesla told him that as a result of his latest experiments he had 'absolutely demonstrated the practicability of the establishment of telegraphic communication to any point of the globe'.[10] He wanted Westinghouse to lend him one of the company's steam-driven dynamos as well as investing in the enterprise and advancing $6,000 against Tesla's royalties for use of his patents in Britain. Westinghouse agreed to lend him the machinery and advance the money, but was unwilling to take the risk of further investment in Tesla's schemes.

Tesla also turned to another former patron, John Jacob Astor. Astor, the son of a fabulously wealthy New York dynasty, was an enthusiast for scientific speculation. He had even written his own scientific romance about life in the year 2000 – *A Journey in Other Worlds: A Romance of the Future*, published in 1894. Astor had been one of the directors of the Cataract Construction Company responsible for the Niagara Falls electrical power project. He had already advanced Tesla $30,000 against shares in the Nikola Tesla Company a few years previously. This time, however, Astor remained unmoved and no further finance was forthcoming from his coffers. Tesla had more luck with another potential investor. J.P. Morgan was the son of a prominent Boston banking family who by the end of the 1890s was one of the most powerful figures in the American financial world.[11] He had already made unsuccessful attempts to buy up Marconi's American patents.

In meetings and in letters Tesla bombarded Morgan with information about his system of wireless telegraphy and the investment that would be needed to get it going. He estimated that building telegraph stations to transmit wirelessly across the Atlantic would cost $100,000 for each station, for example. He assured him that his patents were safe from any challenges by Marconi and that the technology he had developed was far superior to anything that

Marconi had at his disposal. He sent him endorsements of previous inventions by eminent men of science like William Crookes and Lord Kelvin. 'These inventions,' Tesla wrote to Morgan, 'the results attainable only by their means – which now I alone am able to accomplish – in your strong hands, with your consummate knowledge and mastery of business – are worth an incalculable amount of money.'[12] By the end of the year, Morgan had agreed in principle to provide some funding and by March 1901 had handed over $150,000 in exchange for a 51 per cent interest in Tesla's patents.

With Morgan's backing and money in the bank, Tesla was ready to start planning and building the new laboratory which would make his vision a reality. At the beginning of March, the *Western Electrician* reported that Tesla's 'plans for the machinery of wireless telegraphy to signal across the ocean have been completed and a site for a plant selected by Nikola Tesla, and that the project will at once be actively begun.' The article announced that the plant was on the New Jersey coast, and also said that Tesla's agents were on their way to Portugal to select a site there for the receiving station.[13] In fact, Tesla had been in negotiation with a Long Island land speculator named James S. Warden, who had offered him 200 acres of land at Wardenclyffe, 65 miles and an hour and a half by train from New York. Work on building the laboratory, designed by the architect Stanford White, started in September.

Rumours of the move to Wardenclyffe started appearing in the press in August. The laboratory would include 'one of the most complete electrical plants that can be purchased', developing 350 horsepower, and would cost almost $150,000. Tesla was confident that he would soon be 'transmitting commercial messages between Wardenclyffe and Europe without the use of wires or cables'. He complained that he 'would have been sending messages across the ocean without the use of wires by this time if the public were not so hard to convince that it could be done'.[14] It was reported that 'the Tesla factory will be 100 ft. square, supplemented by a "wireless" tower 350 ft. high.'[15] By the end of November, newspapers were reporting that the 'principal building, in which power will be developed, has now been practically completed, and steam boilers and engines are on the spot, being installed as fast as possible'.[16]

There was much speculation about Tesla's tower at Wardenclyffe. The tower, which would be 'the foundation for Mr. Tesla's across-the-world flashes', would be 'octagonal in shape and will be 210 feet high, 100 feet in diameter at the base, narrowing down to 80 feet in diameter at the top'. At the base would be 'a well 120 feet deep', and 'across the bottom of the well will be a series of four tunnels, each to be a 100 feet long.'[17] Tesla had in fact wanted the tower to be 600 feet high, but Morgan had refused to put up the substantial additional funds that would be needed. The final tower was actually 187 feet tall, and the terminal at the top was 68 feet in diameter, weighing 55 tons.[18] A description of the plant written a few years later described the tower as standing 'in lonely grandeur and boldly silhouetted against the sky', a source of 'great satisfaction and of some mystification' to the locals.[19]

In practical terms, what Tesla aimed to achieve at Wardenclyffe was a substantially scaled-up version of his Colorado Springs experiments. The massive semi-spherical terminal on top of the tower was the equivalent of the copper ball he had used there. Its purpose was to store huge quantities of electric charge. Similarly, the well and tunnels at the base of the tower were meant to house the equivalent of the metal plates buried in the ground that he had used to securely earth his Colorado Springs apparatus. Electricity would be generated in the main laboratory building and stepped up to a high voltage to be conducted underground to the magnifying transmitter in the tower. Tesla wanted to use this apparatus to generate an electric wave that would travel through the earth and be reflected back to produce a stationary wave in the earth's crust. It would then be possible, he thought, to pick up signals, or even significant amounts of electrical power, at any point on the earth's surface by using appropriate receiving apparatus.

Wardenclyffe was the realization of Tesla's ambitions. 'We are building for the future,' he told newspapers. His competitors' wireless dreams were made of no more than 'networks of flimsy wire'. Tesla was confident that messages could not be 'transmitted without wires for more than 50 or 60 miles without the use of the principles which I have patented throughout the civilized countries of the world'. When his system was perfected, 'you will be able to put an instrument in your house and talk to anyone who has a similar apparatus anywhere in the country without any metallic or artificial connection.'[20] Locals told reporters about the mysterious flashes of lightning nightly emanating from the Wardenclyffe tower. 'All sorts of lightning were flashed from the tall tower and poles last night,' they reported, and 'the air was filled with blinding streaks of electricity which seemed to shoot off into the dark on some mysterious errand.'[21] As Tesla would not be drawn as to just what he was doing, speculation grew about the nature and scope of his experiments.

Tesla was, moreover, the focus of increasingly sharp criticism in the press. Even the usually loyal *New York Tribune* complained that 'he sometimes appears to be oblivious to some of the uncomplimentary interpretations which are put upon his delays.'[22] A long article in the *Electrical Age* at the beginning of 1903 pointedly listed Tesla's unfulfilled promises. 'Ten years ago,' the article said, 'if public opinion in this country had been required to name the electrician of greatest promise, the answer without doubt would have been "Nikola Tesla".' Now though, 'his name provokes at best a regret that so great a promise should have been unfulfilled.' In those ten

The Tesla Collection

The Wardenclyffe tower.

years, Tesla had gone from being 'the last and greatest of electrical wizards' to being someone who 'has made so many startling announcements and has performed so few of his promises that he is getting to be like the man who called "Wolf! Wolf!" until no one listened to him'.[23]

The Tesla Collection

Inside Tesla's laboratory at Wardenclyffe.

Even before Tesla's Wardenclyffe tower was raised, the whole affair was starting to look like a white elephant. Tesla had undertaken many times in different contexts to send wireless signals across the Atlantic from New York to Paris for the 1900 Exhibition. The Paris Exhibition opened to the public in April and closed in the middle of November without there being any hint of a wireless signal from Tesla. A little over a year later, on 12 December 1901, Marconi succeeded in transmitting the first wireless signal across the Atlantic, from Poldhu in Cornwall to Newfoundland. The *Times* American correspondent reported that 'Signor Marconi authorizes me to announce that he received on Wednesday and Thursday electrical signals at his experimental station here from the station at Poldhu, Cornwall, thus solving the problem of telegraphing across the Atlantic without wires.'[24] Tesla had been pipped at the post. Much of the argument he had been using to persuade men like J.P. Morgan that he was the right horse to back in this race for wireless had lost its power.

Now it was Marconi, not Tesla, who seemed to represent the future of wireless telegraphy. In New York, his successful attempt at transatlantic communication was headline news. Interviewed in the *New York Times*, Tesla's former ally, Thomas Commerford Martin, did not miss the opportunity to turn the knife. 'I am only sorry,' he said, 'that Mr. Tesla, who has given the matter so much thought and experimentation, and to whose initiative so much of the work is due, should not also have been able to accomplish this wonderful feat.'[25] H. Cuthbert Hall, the New York manager of the Marconi Company, interviewed ironically enough at Tesla's usual haunt in the Waldorf Astoria, confirmed that they planned to be in commercial competition with the cable companies as soon as possible: 'Plans were formulated some time ago in anticipation of the successful outcome of Signor Marconi's experiments, but I do not care to make them public just now.'[26] Tesla stayed away when Marconi attended as the guest of honour a few weeks later at the American Institute of Electrical Engineers' annual dinner at the Waldorf Astoria.

Marconi's success placed Tesla and the whole of his Wardenclyffe enterprise in a difficult position. On the one hand, Tesla's goal was far more ambitious than simply signalling wirelessly – he wanted to develop ways of transmitting large quantities of electrical energy through the earth. But on the other hand, the prospect of wireless telegraphy had been central to his appeal to potential investors like J.P. Morgan. When Morgan undertook to finance Tesla to the tune of $150,000, he had done so on the understanding that he was investing in a system of wireless telegraphy.

It now seemed, however, that the rival Marconi system had taken an unassailable lead in the wireless race across the Atlantic. Morgan was slow in delivering the final $50,000 of his promised investment to Tesla, and the completion of the Wardenclyffe tower was delayed as a result. That may well have been a response on his part to Tesla's failure, as he presumably saw it, to fully deliver on his promises.

Despite Tesla's pleas for further investment, no further financial assistance was forthcoming from J.P. Morgan. He tried to interest Morgan in what he called a World Telegraph System, where the latest news would, literally, be sent through the earth so that subscribers could pick it up using their own individual receivers. 'The whole earth is like a brain,' he told his former patron, 'and the capacity of this system is infinite, for the energy received on every few square feet of ground is sufficient to operate an instrument, and the number of devices which can be so actuated is, for all practical purposes infinite.' Tesla proceeded to offer shares in the Nikola Tesla Company to his acquaintances in New York high society, but with little luck. Tesla tried forming a new company, the Tesla Electric and Manufacturing Company, to produce coils for X-ray apparatus. He even tried borrowing money from a bank in Serbia.[27] Tesla considered building another Wardenclyffe at Niagara Falls to transmit its excess energy directly to New York.

Tesla revived his suggestion to Morgan for establishing a system of World Telegraphy in 1904 in an article in the *Electrical World and Engineer*. The idea was to establish a number of plants, each of them 'located near some important center of civilization and the news it receives through any channel will be flashed to all points of the globe'. Subscribers to the system would have a 'cheap and simple device, which might be carried in one's pocket', which would 'record the world's news or such special messages as may be intended for it'. Wardenclyffe was now to be designated as the first of those plants, and 'would have been already completed had it not been for unforeseen delays which, fortunately, have nothing to do with its purely technical features'.[28] According to the *New York Herald*, this was 'a plan to do away with newspapers'. If it worked, 'a man may stand in the middle of the Sahara and by means of an inexpensive instrument so small that it may be carried in a vest pocket receive news of events in New York.'[29]

Despite increasing financial difficulties, Tesla remained confident that Wardenclyffe would pay off. From 'this first power plant, which I have been designing since a long time, I propose to distribute ten thousand horse-power under a tension of one hundred million volts, which I am now able to produce and handle with safety'. The energy would be 'collected all over the globe preferably in small amounts, ranging from a fraction of one to a few horse-power'. It might be used for 'the illumination of isolated homes'. There were 'innumerable devices of all kinds which are either now employed or can be supplied, and by operating them in this manner I may be able to offer a great convenience to the whole world with a plant of no more than ten thousand horse-power'.[30] Two years later, Tesla was in court. A former Wardenclyffe machinist, Frank N. Clark, was suing him for unpaid wages to the tune of $889.25. As the *New York Herald* recorded, when asked by Clark's lawyer if he was 'an inventor and promoter', Tesla replied that 'he was not a promoter and never received a fee, but that he was an inventor and could take rank among the foremost men of the age.'[31] Tesla lost the case – the loss seems emblematic of a broader failure.

PART 5

VISIONS OF TOMORROW

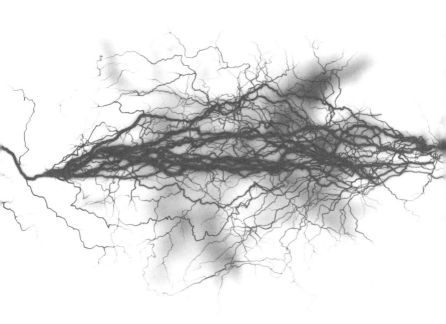

CHAPTER 16

Inventing the Future

In John Jacob Astor's scientific romance, *A Journey in Other Worlds*, the future worked by electricity: 'Electricity in its varied forms does all work, having superseded animal and manual labour in everything, and man has only to direct. The greatest ingenuity next to finding new uses for this almost omnipotent fluid has been displayed in inducing the forces of Nature, and even the sun, to produce it.'[1] The electricity from mountain-top windmills, 'in connection with that obtained by waterfalls, tidal dynamos, thunderstorms, chemical action, and slow-moving quadruple-expansion steam engines, provides the power required to run our electric ships and water-spiders, railways, and stationary and portable motors, for heating the cables laid along the bottom of our canals to prevent their freezing in winter, and for almost every conceivable purpose'.[2] Ships at sea generated their own electricity with wind turbines; electricity flew flying machines that defied gravity and travelled to the stars; electricity controlled the climate and operated kinetographs, or visual telegraphs.

Astor's tale of the future was set in the year 2000, but other futures far closer to Astor and Tesla's present offered a similar story about the coming hegemony of electricity. In Garrett P. Serviss's unauthorized sequel to Wells's *The War of the Worlds*, the fleet of space ships that the fictional Edison built to conquer Mars operated by electricity. The operation of Edison's flying machines 'depended upon the principle of electrical attraction and repulsion'. He had solved the problem of 'how to produce in a limited space, electricity of any desired potential and of any polarity, and that without danger to the experimenter or to the material experimented upon'. Like the space ships in Astor's future, they manipulated electricity to defy gravity. The fleet would be armed with disintegrators. The weapon could 'concentrate its energy upon a given object in order that the atoms composing that object should be set into violent undulation, sufficient to burst it asunder and to scatter its molecules broadcast'.

Closer to earth, the flying machines that featured in George Griffith's various scientific romances of war in the air were operated by electricity and deployed electric weaponry. In novels like *The Angel of the Revolution* (1893) and *The Outlaws of the Air* (1895), Griffith explored the ways in which technological revolution would engender political revolution. In both novels, the flying machines were either invented for or hijacked by anarchist groups who used the immense technological superiority the machines offered to bring the European great powers to their knees. Control of such weaponry meant that the anarchists could bomb cities and countries at will from the air without danger of reprisal. Griffith's short story 'The Raid of Le Vengeur' played out an undeclared war between France and Britain in terms of rival submarine technologies, since 'the nation which could put to sea the first really effective fleet of submarine vessels would hold the fleets of rival nations at its mercy, and acquire the whole ocean and its coasts as an exclusive territory.'[3]

Stories like these offer some background to Tesla's argument that his wireless torpedo boats had the capacity to bring an end to war by depriving powerful nations of their naval superiority. His argument reflected a prevailing assumption about the transformative impact that future technologies might have on political power. These stories are also quite telling in the assumptions they tended to make about just where these kinds of technological innovations came from – and who their inventors were. In Griffith's novels about war in the air, the flying machines that gave their possessors unrivalled power were invented by maverick outsiders. 'VICTORY! It flies! I am master of the Powers of the Air at last!' Those were the words uttered by the flying machine's inventor on the opening page of *The Angel of the Revolution*: 'a pale, haggard, half-starved looking young fellow in a dingy, comfortless room on the top floor of a South London tenement-house; and yet there was a triumphant ring in his voice, and a clear, bright flush on his thin cheeks that spoke at least for his own absolute belief in their truth'.[4]

Authors of scientific romances at the turn of the century populated their tales of the unexpected with misfits and outsiders who had somehow acquired the capacity to remake the future. Inventors were people like Professor James Clinton-Grey in L.J. Beeston's *A Star Fell*, who had discovered 'a new application of the electric fluid whereby I may attain so enormous a speed as to practically annihilate space'.[5] They were people like Arthur Moore, who had invented a perfect 'lady automaton' that could not be distinguished from flesh and blood – and died of a broken heart when his creation was destroyed.[6] If we want to understand what Tesla looked like to his contemporaries – and how Tesla himself thought an inventor should look – then we need to consider his activities in the light of tales like these. Tesla crafted his image carefully and knowingly. He wanted to look like a man who could make the future.

It is worth noting that scientific romances like these filled the pages of magazines like *Pearson's Magazine*, or *Cassell's*, or the *Century*, where articles by or about Tesla himself might also appear. Fictional and factual accounts of the future rubbed shoulders there.[7] Tesla's own essay setting out his manifesto in the *Century Magazine* in 1900 would itself have been understood by many of its readers as a piece of futuristic speculation. Those readers two years earlier would have been reading Tesla's friend Mark Twain's tongue-in-cheek story about the telectroscope a few years in the future.[8] A few months later they would be reading Roger Riordan's musings about 'A Dream of the Future World's Fair', inspired by the Buffalo Pan-American Exposition of 1901. In the Exposition of the Future, the 'clouds themselves will be created and controlled by man ... waste-steam from the machinery will be utilized for the purpose, and artificial rainbows will span the valleys and artificial thunder-storms will sweep the plain'. There would be 'electric suns' to 'illuminate the clouds from within'. In the new century inventors 'like Tesla and Marconi will be given the means to conduct experiments on a large scale'.[9]

H.G. Wells's *Anticipations of the Reaction of Mechanical and Scientific Progress upon Human Life and Thought* first appeared in series form in the *Fortnightly Review* where Crookes had published his speculations about future electricity a decade or so earlier. It was published in the *North American Review* in the United States a little later. This was a concerted effort by Wells to turn his fictional speculations into factual ones and to lay out how future technological change would transform society. It was a given in Wells's argument that the underpinning technology of the future would be electrical. With Niagara presumably in mind, for instance, he imagined how a future Europe would be powered by 'mountain-born electricity ... brought to it in mighty cables from the torrents of the central European mountain mass'.[10] In just the same way that Wells's novels depended on their readers' familiarity with the shape of contemporary technology, his factual futuristic speculations were grounded in the expectation that his readers knew about such things. After all, they were reading about them in the same places they were reading him.

Not just in magazines like these, but in a wide range of other places and publications, speculation about the shape of things to come proliferated. These were boom years for popular science. Books like Charles Gibson's *Romance of Modern Electricity* (1905) went through successive editions and revisions as audiences clamoured for the latest electrical insights into the present and the future. 'If the present conditions of life had been correctly predicted a few generations ago, the prophet would have received little attention, or would have been made a laughing-stock,' he told his readers; but now, the 'present generation, having grown up amidst all these and other wonders, has almost ceased to marvel at them'.[11] The wonderful was becoming commonplace as announcements of new marvels succeeded one another. 'We are surrounded with mysteries, while our vision and all our other powers are limited,' Richard Kerr told the readers of his *Wireless Telegraphy* (1898). There were 'forces and forms of energy, undreamt of, awaiting investigation'.[12]

These kinds of stories, in books and magazines, both fictional and factual, built visions of the future out of the building blocks of the present. Kerr, for example, invited his readers to think about wireless telegraphy in the same breath as wondering about telepathy and oriental mysticism. Wireless was both like and unlike those sorts of mysteries. This sort of blending of fashionable theosophy and technological futurism had a ready market. Louis Pope Gratacap, in his fictional *The Certainty of Future Life in Mars* (1903), had his father and son protagonists speculate whether the spirits of the dead migrated to planets representing different spiritual planes at different historical epochs. At the same time, they were experimenting with wireless telegraphy and hoping to receive signals from Mars. Following his father's death, the son duly received by wireless telegraphy from his dead father, confirming that their theories were correct and that he was now an inhabitant of Mars.

The turn of the century offered particularly rich scope for thinking about the future. With the dawn of a new millennium only another hundred years away, there was much speculation about what the world would look like in the year 2000. That was when Astor's novel was set, for example. Even magazines like the *Ladies' Home Journal* offered their own visions of life a century in the future. In this future, Americans would be two inches taller and live longer, 'Automobiles will be Cheaper than Horses', 'There will be Air-Ships', and 'Aerial War-Ships and Forts on Wheels'. The Americans of the future would attend theatres to see 'the coronations of kings in Europe or the progress of battles in the Orient' thanks to 'cameras connected electrically with screens at opposite ends of circuits, thousands of miles at a span'. Wireless telegraphs, and even wireless telephones, would mean that a 'husband in the middle of the Atlantic will be able to converse with his wife sitting in her boudoir in Chicago'.[13]

As new scientific discoveries and new inventions proliferated, they fed further speculation about the shape of the future. Röntgen's discovery of X-rays generated excited speculation about the possibilities they offered. X-rays would allow people to peer inside the human body, or look through walls. There were other, more prurient possibilities as well:

> I'm full of daze,
> shock and amaze;
> for nowadays
> I hear they'll gaze
> thro' cloak and gown – and even stays,
> these naughty, naughty Roentgen Rays.

Or so suggested an anonymous poet in the journal *Photography*.[14] If X-rays could penetrate flesh and see through solid objects, what might other, undiscovered rays make possible? Tesla himself speculated that X-rays had a particular effect on the mind. After exposing his head to powerful streams of the radiation, he found 'that there is a tendency to sleep and the time seems to pass away quickly'. There was 'a general soothing effect' and 'a sensation of warmth in the upper part of the head'.[15]

Others speculated about other prospects. An article in *Pearson's Magazine* enthused about the researches of the Cambridge-trained Bengali physicist Jagadish Chandra Bose, Professor of Physics at the University of Calcutta, who had experimented on the possibilities of an artificial electric eye that could see the invisible. His electrical eye 'worked on somewhat similar principles to the real eye; there is a sensitive layer on which the invisible light falling gives rise to an electric impulse, which is carried by conducting wire and produces a twitching motion to a part corresponding to the brain'. Similarly, the English medical electrician William Snowdon Hedley suggested that eyes might be imagined as wireless receivers, 'syntonised for the reception of similarly vibrating etherial impulses radiating from some given source'.[16] Bose speculated about the further uses of the ether waves his electrical eye detected. Ships at sea might be equipped with electric eyes that would allow them to see signals from electric lighthouses warning them of danger.[17] Bose, who was also an author of scientific romance, filed an application for a US patent for his electrical eye, or 'detector for electrical disturbances' in 1901.[18]

The discovery of radioactivity sparked similar speculation. Just as Röntgen at the end of 1895 had been intrigued by the cause of fogged photographic plates, so had Henri Becquerel. In an announcement to the French Academy of Sciences in February 1896 Becquerel described how a mysterious radiation emanating from uranium salts caused photographic plates to become foggy, even when wrapped in black paper that light could not penetrate.

Becquerel's work was taken up by Marie Curie and her husband Pierre, who in 1898 announced that they had identified two hitherto unknown elements that seemed to give off this mysterious radiation too. They named the elements polonium and radium and invented a new word to describe their condition – they were radio-active. The *New York Times* reported William Crookes as speculating that 'if half a kilogram were in a bottle on that table it would kill us all.' It generated such energy that 'one gram is enough to lift the whole of the British fleet to the top of Ben Nevis; and I am not quite certain that we could not throw in the French fleet as well.'[19]

Tesla was less impressed by Becquerel and the Curies. 'I do not believe that this discovery will be useful in wireless telegraphy,' he told the *New York Journal and Advertiser*. In fact, he thought that 'telegraphy with rays in general is of little practical value.' As far as he was concerned the discovery of radioactivity was another dead end – like Marconi's 'Hertzian telegraphy'.[20] Others imagined a future in which the health-giving rays of radioactivity would banish disease. Like X-rays, radioactivity fuelled speculation about other kinds of rays that might be discovered in the future. For a brief period between 1903 and 1904 there was excited speculation about the properties of the new N-rays, discovered by the French physicist René Blondlot. They were particularly intriguing since it seemed that they were generated by the nervous systems of living things. Another French physicist, Arsène d'Arsonval, demonstrated that the mysterious rays seemed to be given off by Broca's centre in the brain during speech. Unfortunately for Blondlot and d'Arsonval, by the end of 1904 it had become clear that N-rays were a figment of the imagination.

N-rays were an example of just how fluid the future might be at the beginning of the twentieth century. If X-rays could penetrate through solid objects, or radioactivity demonstrate the enormous energies locked up in apparently ordinary matter, then there was nothing implausible about strange rays emanating from the human brain. It was just the sort of thing that William Crookes had speculated about at the beginning of the 1890s. His predictions of wireless telegraphy had turned out to be accurate. Why not his predictions about electric organs in the brain? Similarly, if Marconi had shown it was possible to send wireless messages across the Atlantic Ocean, why not assume with Tesla that wireless messages could be sent across the ocean of ether between the earth and Mars as well? It was through this sort of speculation – playing with the boundaries of the possible through fact and fiction – that the future was being built at the beginning of the twentieth century. Tesla was one of its chief architects.

CHAPTER 17

Projections

Despite his dogged persistence and inventiveness, by the end of the first decade of the twentieth century Tesla's vision of the future was starting to look less and less like the one that was actually unfolding around him. Despite his best efforts, the work at Wardenclyffe appeared to be going nowhere, and his debts were mounting. At the beginning of 1906, equipment remaining at his Colorado Springs laboratory was seized to pay an outstanding debt.[1] Tesla continued to maintain to the press the merits of his system of wireless telegraphy and its superiority to Marconi's 'primitive Hertzwave signalling'. 'The fact is my Long Island plant will transmit almost its entire energy to the antipodes if desired,' he wrote to the newspapers.[2] But even as Tesla continued to insist that Wardenclyffe was still a going concern, there were rumours that he was in serious financial difficulty. Only a few days after they published his latest letter, the *New York Times* reported that the Tesla tower was for sale – and not for the first time. It too had been seized by the local sheriff to be sold to pay an outstanding debt.[3]

On that occasion Tesla managed to scrape together the funds he needed to save the tower, but even as one debt was paid off, others were mounting. By 1904 Tesla had already mortgaged Wardenclyffe to George Boldt, the proprietor of the Waldorf Astoria hotel, in order to pay his mounting hotel bills. In 1917 the tower itself was torn down and sold for scrap to pay off creditors.[4] Even as his future seemed to be collapsing around him, Tesla was continuing to speculate. At the beginning of 1908 the *New York World* posted his predictions for the coming year. 1908 would 'mark the end of a number of erroneous ideas which, by their paralyzing effect on the mind, have throttled independent research and hampered progress and development in various departments of science and engineering'. The 'illusion of the Hertz or electro-magnetic waves', would become clear to everybody; it would become clear too that 'there is no such element as radium, pollonium [*sic*] or ionium' – radioactive emanations were 'emitted more or less by all bodies, and are all of the same kind'. Tesla predicted that aeroplanes would 'never fly as fast as a dirigible balloon'.[5]

Tesla was – presumably – trying to be provocative, and keep himself in the public's mind. Despite the setbacks, he was still projecting his visions of the future, and would keep on doing so. Tesla did not relinquish his dream of wireless electrical energy transmitted through the earth, even if by 1920 the means of realizing that dream had been dismantled. Interviewed by the *New York Herald* in 1912, he was still complaining that the 'sensational exploitation of wireless telegraphy and telephony has naturally created in the public mind the idea that this is the chief if not the only field of its use'. As a result, 'the world has not even at this moment the faintest conception of the really great and valuable results in other fields that are sure to be obtained in the near future.' The 'transmission of power wirelessly' would be the really transformative achievement and would open up a world of possibilities. How about 'the construction of one single plant from which all the flying machines of the world could be operated without fuel or other energy of any kind!'?[6]

A year later, in an interview with the *New York Press*, Tesla was still adamant that the 'wireless art' was the key to the future. He was sure that 'its application for innumerable purposes to which it lends itself will be of greater significance in our future development than any other technical advance.' It would be 'the means of completely annihilating distance and thereby affording unprecedented opportunities for intellectual and material advancement'. His wireless system would revolutionize 'the three great departments of human activity' – the 'transmission of intelligence', the 'transport of our bodies and materials', and the 'transmission of energy necessary to our existence'. The first was already happening and as expertise and knowledge grew he was confident that the other two would follow. When the problem of wirelessly transmitting energy was solved, 'cheap power will become available at any place in the world', and 'the sinful waste of energy which is now going on in this country perhaps more so than in any other' would come to an end.[7]

In an article on 'Famous Scientific Illusions' in 1919 Tesla included what he called 'the singular misconception of the wireless'. The notion that wireless operated by means of waves transmitted through the atmosphere was a 'monstrous idea'. Wireless travelled through the earth. The only result of experts' obsession with Hertzian waves was that 'the true wireless art, to which I laid the foundation in 1893, has been retarded in its development for twenty years'. Tesla remained a true believer. He was still confident that it would be possible to 'generate electrical energy at the source of supply, Niagara, Kaieteur or Rjukan, for instance, and transmit it by wireless to the nearest receiving stations in other countries'. It would not be 'scattered promiscuously to the four winds of heaven for the free use of all' but protected by 'a secret key or combination like that of a safe, so that only those for whom the power is intended will get it'.[8]

Just as he had done with his earlier plans for a wireless torpedo, Tesla continued to ponder how his system could be used to develop weapons so terrifying that they would render war unthinkable. In an article written for the *New York World* in 1907 he speculated about how his wireless torpedo might be used to generate a tidal wave so huge that it could wipe out entire fleets of battleships. The new class of dreadnaughts being developed in Europe would be helpless against this kind of attack. Tesla imagined one of his small, manoeuvrable torpedo boats, loaded with 'twenty or thirty tons of cheap explosive'. The whole thing would be 'under the perfect control of a skilled operator far away'. The explosion would lift 25,000,000 tons of sea water into the air, he estimated, and the 'entire navy of a great country, if massed around, would be destroyed'. Even a dreadnaught some distance away would be thrown into the air and then 'sink far below the surface, never to rise'.[9]

Tesla was pessimistic about the prospects of heavier-than-air flight, despite the successes of the Wright brothers. As far as he was concerned, success would depend on 'principles radically novel'.[10] Conventional engines could never be made light enough for powered flight on a large scale. His solution to the problem was to invent a new kind of engine. 'Ten horsepower from a tiny engine that a man could dangle from his little finger by a string! Five hundred horsepower in a package that a man could lift easily in one hand! A thousand horsepower motor occupying hardly more space than the cardboard box in which your hatter sent your new derby home!' – that was what Tesla was now offering. It was 'what mechanical engineers have been dreaming about ever since the invention of steam power', he claimed. It was 'the perfect rotary engine'.[11] He was confident 'that there was no limit to the capacity for which this turbine could be built'.[12]

This was the engine that Tesla thought could realize the dream of powered flight. This was 'the object towards which I have been directing my energies for more than twenty years – the dream of my life,' he now claimed, though there is little evidence to suggest that achieving powered flight was much on his mind during the 1890s. But Tesla was not above rewriting his own past as he fought desperately to keep up with the changing future, and insisted that while he had been working away at wireless transmission, his real goal had been 'a flying machine propelled by an electric motor, with power supplied from stations on the earth'. Now he had turned his attention to mechanical engines, and this was the result. Rather than being used to power mere aeroplanes, which could never be anything but 'a toy – a sporting plaything', his engines would be used to power real flying machines. These flying machines of the future would 'have neither wings nor propellers'. Seen on the ground they would not even look like flying machines. They would 'be able to move at will through the air in any direction with perfect safety, higher speeds than have yet been reached, regardless of weather and oblivious of "holes in the air" or downward currents'.[13]

Tesla – unsurprisingly for an electrical experimenter who had established his public reputation by passing massive currents through his own body – had a longstanding interest in the therapeutic uses of electricity. Middle-class America at the turn of the new century both celebrated and worried about nervousness. Nervousness was a symptom of a progressive industrial culture, but allowed to get out of control it could lead to personal and cultural degeneration as well. In 1898 Tesla had delivered a paper to the American Electro-Therapeutic Association on the possible medical uses of his high-frequency oscillator. The oscillators were particularly useful for electrotherapy because they could deliver particularly high doses of electricity without killing the patient. They could be used to 'pass through the body, or any parts of the same, currents of comparatively large volume under a small electrical pressure', or to 'subject the body to a high electrical pressure while the current is negligibly small', or even 'to put the patient under the influence of electric waves'.[14]

Minds as well as bodies might benefit from judicious doses of electricity. In 1912, at Tesla's suggestion, a New York school embarked on an experiment in electrical education. The idea was to investigate whether electricity could make children cleverer, and 'lessening the burden of school life and the difficulties of acquiring education'. At the start of each school day 'the high frequency current will be turned on, by one of the instructors, and shortly afterwards the school room will become completely saturated with infinitesimal electrical waves vibrating at high frequency.' The entire room would be turned into one big 'health-giving and stimulating electro-magnetic field, or "bath"'. The process worked by a kind of 'molecular massage or tissue gymnastics', as the current caused the 'tiny particles of which the body is composed' to 'move about in a livelier fashion and increase in number'. The proposal was based on a plan that had already been carried out in Stockholm by the physical chemist Svante Arrhenius.[15]

The plan for 'stimulating dull pupils by saturating them with electricity' was actually based on some old experiments by Oliver Lodge, to investigate whether plants could be stimulated into growing more quickly by being exposed to a powerful electro-magnetic field. A new potential use for this apparatus was in the family home, said Tesla. The 'up to date home of the near future' would be a 'health resort and sanatorium as well, making sea side vacations unnecessary'. They would come equipped with his high-frequency electrical apparatus so that the entire family could be kept constantly charged with electricity, making them 'fit for the battle of life'. Thanks to Tesla's apparatus, 'all the benefits of the seashore may be obtained right in the crowded city.' Several years later, he was still advocating 'Tesla currents' that would 'by softening the arteries, make the old feel younger, and the young, younger, and more aggressive'. 'Women, in particular, should derive great benefit from their future high frequency dry baths,' he told the pioneer of science fiction magazines Hugo Gernsback.[16]

By the 1920s though, Tesla was finding it more difficult to cultivate his image as the man of the future. His financial problems continued to get worse. Hugo Gernsback helped out by publishing his autobiography in instalments in *Electrical Experimenter* in 1919, and later persuaded the Westinghouse Company to offer him a monthly pension. He had long since abandoned the opulent Waldorf Astoria for more modest surroundings. None of this halted the flow of speculation though. In 1931 as he prepared for his 75th birthday, Tesla announced again that he was on the brink of announcing the discovery of a revolutionary new source of energy. His discovery would 'make it possible to tap inexhaustible streams of power at any point on the globe', and 'great changes will follow in social life'.[17] He was coy about just what this new source of energy was, though insistent that he would reveal all once his experiments were finished. Newspapers speculated: was it the sun's energy, was it energy from space? Tesla was emphatic that whatever it was, it was not atomic energy, which he condemned as a chimera.

The stream of speculation continued. He had been 'experimenting with rocket-propelled ships' and anticipated that 'such machines will be of tremendous importance in international conflicts of the future.'[18] He had not forgotten Mars either. He announced that 'he had developed after many years of concentrated effort a means that will make it possible for man to transmit energy in large amounts, thousands of horse power, from one planet to another, which, he believes, some day will open the way for interplanetary communication.'[19] On his 77th birthday in 1933 he repeated his claim that he was on the verge of disclosing a revolutionary new source of energy. Once built, he said, the machines providing the energy would last 500 years.[20] The discovery was so fundamental that 'it will undo the Einstein theory of relativity', and the machines producing the energy would be simpler than 'any machines ever invented for the production of power'.[21] Tesla in his seventies was enjoying another flurry of publicity. The *New York Sun* called him 'the Jules Verne of modern science'.[22]

Tesla was making good use of the newspaper interviews that coincided each year with his birthday to revive old enthusiasms. Wireless energy, interplanetary communication, and new designs of flying machine were all revisited. In 1934 it was the turn of his old preoccupation with weaponry that could end all wars. On his 78th birthday he announced the invention of 'a beam of force somewhat similar to the death ray of scientific romance'. The weapon would be capable of 'destroying an army 200 miles away', it could 'bring down an airplane like a duck on the wing', and it could 'penetrate all but the most enormous thicknesses of armor plate'. The ray would 'be generated at stationary power plants by machines which involve four electrical devices of the most revolutionary sort'. It was to be a weapon of peace, and in peacetime the weapon would be used 'to transmit immense voltages of power over distances limited only by the curvature of the earth'.[23]

Throughout the final decade of his life, as first Europe and then America became embroiled in total war, Tesla kept on dreaming his electric dreams, and hoping for the investment that would make them a reality. There is little to suggest that his schemes ever progressed beyond the drawing board though – or even that they progressed as far as that. They mainly existed inside Tesla's head.

Tesla died in his sleep on 7 January 1943, at the Hotel New Yorker where he had been living for some years. He left behind him unpaid debts and rooms full of paper. The papers – particularly in view of his speculations about death rays – were of sufficient concern for them to be confiscated by the FBI and closely studied, though the researchers soon concluded that they contained nothing of interest. His funeral took place a few days later at the Cathedral

of St John the Divine in New York. 'Inventors, Nobel Prize win-
ners, leaders in the electrical arts, high officials of the Yugoslav
Government and of New York, and men and women who attained
distinction in many other fields' were among the 2,000 mourners.[24]

After the end of the war, Tesla's nephew Sava Kosanović, who
had spent some of the war years in America as a supporter of the
exiled Yugoslav King Peter II, returned to the United States as
the Yugoslav ambassador. He paid off Tesla's debts and took the
papers back with him to Belgrade where they would become part
of a museum dedicated to Tesla. But, of course, Tesla left behind
far more than debts and papers. He left behind him a whole new
vision of the future.

CHAPTER 18

The Afterlives
of Nikola Tesla

Nikola Tesla did not believe in an afterlife, though he was sufficiently impressed by William Crookes's researches into spiritualism to take the idea seriously. For a brief period following his mother's death he even thought that he might have had a spiritualist experience himself, but soon decided that the experience had a more mundane explanation. In interviews towards the end of his life he confirmed that he was an advocate of what he called 'the mechanistic conception of life'. As far as he was concerned, 'what we call "soul" or "spirit" is nothing more than the sum of the functionings of the body.' Accordingly, when 'this function ceases, the "soul" or the "spirit" ceases likewise'.[1] His lack of belief in an afterlife did not stop him from wishing for immortality though. Tesla certainly wanted to live forever, and he was confident that his inventions would make sure that he did. He had 'been wonderfully fortunate', he thought, 'in the evolution of new ideas, and the thought that a number of them will be remembered by posterity makes me happy indeed'.[2]

Tesla's dreams in that respect have certainly been realized. Posterity remembers Nikola Tesla. Since his death in 1943 he has been the subject of dozens of biographies. The first of them, *Prodigal Genius: The Life of Nikola Tesla*, by the Pulitzer Prize-winning journalist John O'Neill, was published the year following his death.[3] O'Neill, as a journalist for the *New York Herald Tribune*, had known Tesla personally, and his biography in many ways set the tone for those that followed. The biography received a rapturous review in *Time*, quoting O'Neill's description of his hero as 'a superman – unquestionably one of the world's greatest geniuses'.[4] As well as books, Tesla has featured in any number of magazine and journal articles in the 76 years since his death. In the former Yugoslavia and in present-day Serbia, Tesla was and remains a national hero. There is a Nikola Tesla Museum in Belgrade. In 1960 the name Tesla was adopted for the unit of magnetic induction, or flux density, in the SI international system of scientific units. His place in the history of science seems assured.

Tesla occupies a prominent place in contemporary popular culture too. In 2006 David Bowie played him in the movie *The Prestige*. He appears as a character in numerous steampunk novels. In 2003 the internet entrepreneur and co-founder of PayPal, Elon Musk, named the company he established to design and manufacture electric cars Tesla. Famously, in 2018 another of Elon Musk's companies, SpaceX, launched a 2008-model Tesla Roadster into space on top of one of their Falcon rockets, and into an orbit that will eventually intersect that of Mars. The original Tesla would presumably have been amused. Sheldon Cooper, one of the main characters in the long-running comedy series *The Big Bang Theory*, makes frequent references to Tesla and treats him as one of his heroes. In view of all this, it is an oddity of many contemporary discussions of Tesla that they are presented as rescue missions. Tesla, the icon of popular and scientific culture is a forgotten genius, badly in need of recovering from obscurity.

In many ways, this is exactly as Tesla would have wanted it. The image of the neglected outsider was one that he carefully cultivated, particularly in later life. It suited Tesla to present himself – on occasion at least – as someone who worked outside the usual conventions, though there were clearly other times when he craved the recognition of his professional peers. At one extreme, the failure of some of Tesla's schemes has been interpreted as evidence of a conspiracy. Some contemporary Tesla enthusiasts believe that the Wardenclyffe experiments were attempts to generate free energy, either from the earth or from the atmosphere. From that perspective, J.P. Morgan's refusal to continue to finance Tesla's efforts, for example, has been portrayed as a piece of deliberate sabotage by a robber baron who feared the threat that endless supplies of free energy posed to the economic system that had enriched him.

Tesla's announcement of his invention of a death ray a few years before his death has attracted similar attention. The death ray was real, from this perspective, and the FBI's seizure of Tesla's papers after his death offers proof that the weapon really existed. The fact that Tesla was himself – as he often was in announcements regarding his latest inventions – deliberately vague about the details, has been taken as straightforward evidence that he really did have something to hide. Some of the theories surrounding Tesla have an air of surreality. His accounts of the possibility of interplanetary communication have been taken as evidence of actual alien encounters. According to some, Tesla was himself an alien, deposited at birth on Djuka and Milutin's doorstep in 1856. The Wardenclyffe tower has been described as shaped like a pyramid, having been designed, like the Egyptian pyramids themselves, to draw energy from the atmosphere. What is it about Tesla that makes him a focus for this kind of speculation?

The ways in which the rivalry between Tesla and Edison is often portrayed is particularly fascinating – and revealing – in this respect. In fact, it is particularly revealing that the relationship between the two is so often portrayed as rivalry at all. Looked at from one point of view, after all, the interaction between the two men was fairly minimal. Tesla had worked for one of Edison's European companies in Paris for a couple of years before emigrating to the United States. There, he worked for the Edison Machine Works for a few months before resigning and setting out on his own as an electrical engineer. Beyond a few possibly apocryphal tales of encounters between the two, it is not even clear how often they met. Despite this, the fierce battle of the systems that was waged between proponents of alternating and direct current electrical systems during the 1880s and early 1890s is often framed as a bitter rivalry between the two inventors.

In reality, Tesla has little directly to do with the battle of the systems. By the time he became a figure to reckon with in the world of American electrical engineering, following his triumphant lecture to the American Institute of Electrical Engineers in 1888, the battle lines were already drawn, and they were between Edison and Westinghouse, not Tesla. It seems unlikely that Edison regarded any relationship with Tesla as a rivalry, though Tesla may have viewed things differently. It does seem clear that Tesla to some extent modelled himself on the older electrician. He was certainly alert to the extent to which Edison worked his magic on the American public through spectacle and showmanship. From the 1890s onwards, he developed the same sort of approach to publicity himself – even if the image he portrayed was rather a different one from Edison's public persona of rugged, self-made, practical man. Like Edison, he was well aware of the value of providing good copy for newspapers.

The rivalry between Edison and Tesla is very much the product of the two men's afterlives. Edison has come to occupy a very particular place in the American cultural imagination. Tesla, by contrast, is the anti-Edison. Edison made himself into the consummate self-made man who remade America with electricity. An old episode of *The Simpsons* from 1998 ('The Wizard of Evergreen Terrace') makes the point, with Homer depressed as he compares his feckless self to the tireless inventive genius who produced the technologies and systems that made the modern world: the phonograph, the incandescent light bulb, the kinetoscope – the list could go on. From this perspective Edison was both the lone wolf inventor who did it all himself, and the corporate giant who founded General Electric. In these sorts of stories, Tesla gets to represent the might-have-been – the alternative America that might exist now if he, rather than Edison, had come out on top in that imaginary rivalry.

One way of responding to Edison's hype is to turn him into a villain instead of a hero. Rather than being the heroic inventor of the future, Edison gets cast as the con artist – the man who manipulated the media to get the credit for achievements that were not really his after all. Edison gets to carry the can for an energy distribution system seen to benefit producers at the expense of consumers. In that kind of story Tesla plays the role of the heroic and neglected underdog. Where Edison was a ruthless exploiter of others' inventions for his own personal gain, Tesla gave his inventions away for little or nothing. Where Edison invented to make money, Tesla invented for the sake of invention. It is easy to see Tesla in this respect as the inventor of an alternative future of the twentieth century where technological innovation was somehow decoupled from corporate greed. What if Tesla had won – would we now live in a world of boundless free energy and flying machines powered by wireless?

In reality though, Tesla was as much an inventor of our present as Edison was. It is a pervasive feature of the way we tend to think about inventions and inventors now that we think about them in the singular. Inventions are invented by great individuals. This way of thinking about invention is deeply ingrained in our culture. Tesla was one of the people who made it so. It was during the nineteenth century that people started regarding inventors as heroic individuals.

The trend had started well before Tesla, of course – think of James Watt, Isambard Kingdom Brunel, or Samuel Morse and the ways they were made into heroes of the industrial revolution – but Tesla (like Edison) was just as keen as anyone to play with this portrayal of the heroic inventor. It was in this context that Tesla, along with his patrons and the newspapers anxious to fill their pages with titillating sensation, helped create an image of the inventor that continues to resonate today.

This way of thinking about inventions and inventors has important implications for the way we think about the future as well. The future as a place that is different from the present is very much a nineteenth-century invention too. During the nineteenth century people started imagining futures where societies would be differently organized and where new technologies would change the ways in which lives were lived in ways they had not done before. International exhibitions offered glimpses of what that future world would look like, populated by technologies that were at once both strange and familiar. Books and magazines offered glimpses of the future too – through both fact and fiction. Speculation about tomorrow and its technologies moved to and fro between the speculative fiction of writers like Jules Verne and H.G. Wells and the promises made by inventors like Tesla and Edison about the new worlds their technologies could deliver. The way that Tesla talked about the future at the beginning of the twentieth century helped forge the way we talk about it now at the beginning of the twenty-first century.

A key feature of Tesla's talk about the future (just like his talk about invention) was that making the future was a matter for extraordinary individuals. Tesla very clearly saw himself as someone who had the ability to remake the future in his own image. He had it all mapped out – and it was a future that depended on him.

Tesla painted a future of limitless (not free) energy beamed effort-lessly through the earth; of flying machines drawing their power from that universal source; of wireless weapons that would end war; and interplanetary communication. He also painted himself as the sole author of that future. The future was going to belong to him. This was – and is – a very powerful image of how the future can be forged. It is a measure of just what a good storyteller about future worlds Tesla was that we still find the story so compelling. It is also the way we still tend to tell stories about imagined futures now. We still tend to frame the way we think about scientific and technologi-cal innovation – the things on which our futures will depend – in terms of the interventions of heroic individuals battling against the odds. A hundred years after Tesla, it might be time to start thinking about other ways of talking about the shape of things to come and who is responsible for shaping them.

Bibliography

Books

Alden, William Livingston, *Van Wagener's Ways* (London: C. Arthur Pearson Ltd, 1898)

Astor, John Jacob, *A Journey in Other Worlds: A Romance of the Future* (New York: D. Appleton and Co., 1894).

Baker, E.C., *Sir William Preece F.R.S.: Victorian Engineer Extraordinary* (London: Hutchinson, 1976)

Beard, George, *American Nervousness* (New York: G.P. Putnam's Sons, 1881)

Beauchamp, K.G., *Exhibiting Electricity* (London: Institution of Electrical Engineers, 1997)

Bowler, Peter, *The Invention of Progress* (Oxford: Blackwells, 1989)

Brown, Harold P., *The Comparative Danger to Life of the Alternating and Continuous Electrical Currents* (New York, 1889).

Cahan, David, *An Institute for an Empire: The Physikalisch-Technische Reichsanstalt 1871–1918* (Cambridge: Cambridge University Press, 1989)

Carlson, W. Bernard, *Tesla: Inventor of the Electrical Age* (Princeton: Princeton University Press, 2013)

Caufield, Catherine, *Multiple Exposures: Chronicles of the Radiation Age* (Harmondsworth: Penguin, 1989)

Chernow, Ron, *The House of Morgan: An American Banking Dynasty and the Rise of Modern Finance* (New York: Grove Press, 2010)

Clarke, I.F. (ed.), *The Tale of the Next Great War, 1871–1914* (Liverpool: Liverpool University Press, 1995)

Cook, James, *The Arts of Deception* (Cambridge MA: Harvard University Press, 2001)

Davenport, Walter Rice, *Thomas Davenport: Pioneer Inventor* (Montpelier: Vermont Historical Society, 1929)

Dyer, Frank and Thomas Commerford Martin, *Edison: His Life and Inventions* (New York: Harper & Brothers, 1910)

Essig, Mark, *Edison and the Electric Chair* (Stroud: Sutton Publishing, 2003)

Fahie, J.J., *A History of Wireless Telegraphy* (London: Blackwoods, 1899)

Gibson, Charles, *The Romance of Modern Electricity*, revised edition (Philadelphia: J.B. Lippincott, 1910)

Gooday, Graeme, *Domesticating Electricity: Technology, Uncertainty and Gender, 1880–1914* (London: Routledge, 2008)

Griffith, George, *The Angel of the Revolution: A Tale of the Coming Terror* (London: Tower Publishing Company, 1893)

Grindriez, Charles and James M. Hart, *International Exhibitions: Paris–Philadelphia–Vienna* (New York: A.S. Barnes & Co., 1878)

Grove, William Robert, *The Correlation of Physical Forces*, 6th edn (London: Longmans, Green and Co., 1874)

Headrick, Daniel, *The Invisible Weapon: Telecommunications and National Politics, 1851–1945* (Oxford: Oxford University Press, 1991)

Hubbard, Geoffrey, *Cooke, Wheatstone and the Invention of the Electric Telegraph* (London: Routledge & Kegan Paul, 1965)

Hughes, Ivor and David Ellis Evans, *Before We Went Wireless: David Edward Hughes FRS, His Life, Inventions and Discoveries* (Benington VT: Images from the Past, 2011).

Hughes, Thomas, *Networks of Power: Electrification in Western Society, 1880–1930* (Baltimore MD: Johns Hopkins University Press, 1983)

Israel, Paul, *Edison: A Life of Invention* (New York: John Wiley, 1998)

Judson, Pieter, *The Habsburg Empire: A New History* (Cambridge MA: Harvard University Press, 2016)

Kerr, Richard, *Wireless Telegraphy, Popularly Explained* (New York: Charles Scribner's Sons, 1898)

Kieve, Jeffrey, *The Electric Telegraph: A Social and Economic History* (Newton Abbott: David & Charles, 1973)

Lodge, Oliver, *Modern Views of Electricity* (London: Macmillan, 1889)

Lodge, Oliver, *Modern Views of Electricity*, 2nd edn (London: Macmillan, 1892)

Lodge, Oliver, *Past Years: An Autobiography* (London: Hodder & Stoughton, 1931)

Lytton, Edward Bulwer, *The Coming Race* (London: Blackwoods, 1871).

Martin, Thomas Commerford (ed.), *The Inventions, Researches and Writings of Nikola Tesla*, 2nd edn (New York: Barnes & Noble, 1995)

Marvin, Carolyn, *When Old Technologies were New: Thinking About Electrical Communication in the Late Nineteenth Century* (Oxford: Oxford University Press, 1988)

Maxwell, James Clerk, *A Treatise on Electricity and Magnetism*, 2 vols (Cambridge: Cambridge University Press, 1873)

McGreevy, Patrick, *Imagining Niagara: The Meaning and Making of Niagara Falls* (Amherst: University of Massachusetts Press, 2009)

Mom, Gijs, *The Electric Vehicle: Technology and Expectations in the Automobile Age* (Baltimore MD: Johns Hopkins University Press, 2004)

Morse, Edward Lind, *Samuel F.B. Morse: His Letters and Journals*, 2 volumes (Boston: Houghton Mifflin, 1914)

Morus, Iwan Rhys, *Frankenstein's Children: Electricity, Exhibition and Experiment in Early Nineteenth-Century London* (Princeton: Princeton University Press, 1998)

Morus, Iwan Rhys, *Michael Faraday and the Electrical Century* (London: Icon Books, 2005)

Morus, Iwan Rhys, *When Physics Became King* (Chicago: University of Chicago Press, 2005)

Morus, Iwan Rhys, *Shocking Bodies: Life, Death and Electricity in Victorian England* (Stroud: History Press, 2011)

Morus, Iwan Rhys, *William Robert Grove: Victorian Gentleman of Science* (Cardiff: University of Wales Press, 2017)

Moyer, Albert, *Joseph Henry: The Rise of an American Scientist* (Washington DC: Smithsonian Institution Press, 1997)

Mrkich, Dan, *Tesla: The European Years* (Ottowa: Commoners Publishing, 2010)

Noad, Henry M., *A Course of Eight Lectures on Electricity, Galvanism, Magnetism, and Electro-magnetism* (London, 1839)

O'Neill, John, *Prodigal Genius: The Life of Nikola Tesla* (New York: Ives Washburn, 1944)

Pole, William, *The Life of Sir William Siemens* (London: John Murray, 1888)

Post, Robert, *Physics, Patents and Politics: A Biography of Charles Grafton Page* (New York: Science History Publications, 1976)

Reports of the United States Commissioners to the Universal Exposition of 1889 at Paris (Washington DC: Government Printing Office, 1891)

Robida, Albert, *The Twentieth Century*, trans. Philippe Willems (Middletown CT: Wesleyan University Press, 2004)

Schivelbusch, Wolfgang, *Disenchanted Night: The Industrialization of Light* (Berkeley, London and Los Angeles: University of California Press, 1988)

Science Fiction by the Rivals of H.G. Wells (Secaucus, NJ: Castle Books, 1979)

Sellers, Charles Coleman, *Mr. Peale's Museum: Charles Willson Peale and the First Popular Museum of Natural Science and Art* (New York: Norton, 1980)

Shorter, Clement, *The Brontës: Life and Letters* (London: Hodder & Stoughton, 1908)

Smee, Alfred, *Elements of Electrometallurgy*, 3rd edn (London, 1844)

Smith, Crosbie and M. Norton Wise, *Energy and Empire: A Biographical Study of Lord Kelvin* (Cambridge: Cambridge University Press, 1989)

Tattersdill, Will, *Science, Fiction, and the Fin-de-Siècle Periodical Press* (Cambridge: Cambridge University Press, 2016)

Tesla, Nikola, *My Inventions and Other Writings* (London: Penguin Books, 2011)

Thompson, Silvanus P., *Life of Lord Kelvin*, 2 vols (London: Macmillan, 1910)

Verne, Jules, *Twenty Thousand Leagues Under the Sea*, first published 1869, this edn trans. Mendor T. Brunetti (Harmondsworth: Penguin, 1994)

Weeden, Brenda, *The Education of the Eye: History of the Polytechnic Institution, 1838–1881* (Cambridge, UK, 2008)

Wells, H.G., *Anticipations of the Reaction of Mechanical and Scientific Progress upon Human Life and Thought* (London: Chapman & Hall, 1902)

White, Trumbull and William Ingleheart, *The World's Columbian Exposition* (Boston: John K. Hastings, 1893)

Articles

'2,000 Are Present at Tesla Funeral', *New York Times*, 13 January 1943

'500,000 Volts of Electricity Passed Through Body to Cure Consumption – Tesla's Idea', *New York World*, 19 August 1900

'A Hungarian Wizard', *Pall Mall Gazette*, 22 June 1900

'A Submarine Destroyer that Really Destroys', *New York Journal*,
        13 November 1898

'A Tesla Patent', *Electrical World and Engineer*, 31 March 1900

'A Town Lighted by Electricity', *Scientific American*, May 1880

'Astounding News!' *New York Sun*, 13 April 1844

'At the Fair', *Century Magazine*, 1893, 46: 3–21

'Beam to Kill Army at 200 Miles', *New York Herald Tribune*, 11 July 1934

Callan, Nicholas, 'On a New Galvanic Battery', *Philosophical Magazine*,
        1836, 9: 472–78

Carey, George R., 'Seeing by Electricity', *Scientific American*, 1880, 42: 355.

Carlson, W. Bernard, 'Elihu Thomson: Man of Many Facets', *IEEE
        Spectrum*, 1983, 20: 72–75

Carper, Steve, 'The Mighty Electrical Men', https://www.blackgate.
        com/2018/02/07/the-mighty-electric-men/ (accessed 29/02/2019)

Chrisman, Francis Leon, 'Our Modern Franklin – Nikola Tesla', *Success*,
        4 November 1898

Clark, Latimer, 'Inaugural Address', *Journal of the Society of Telegraph
        Engineers*, 1875, 4: 1–22

'Cloudborn Electric Wavelets to Encircle the Globe', *New York Times*,
        27 March 1904

Crookes, William, 'Electricity in Relation to Science', *Popular Science
        Monthly*, 1892, 40: 497–500

Crookes, William, 'Some Possibilities of Electricity', *Fortnightly Review*,
        1892, 51: 173–81

'Crystal Palace – Electrical Exhibition', *Morning Post*, 19 April 1892

'Davenport's Electro-magnetic Engine', *Mechanics' Magazine*, 1837, 27:
        404–5

'Doubts Value of Tesla Discovery', *New York Herald*, 9 November 1898

Editorial, *Electrician*, 1900, 45: 274

Editorial, *New York Sun*, 14 March 1895

Editorial, *Western Electrician*, 9 March 1901

Edwards, E. Jay, 'The Capture of Niagara', *McClure's Magazine*, 1894, 3:
        423–35

'Electric Light News', *Electrical Review*, 16 October 1886

'Electric Light', *Patent Journal*, 1849, 6: 80

'Electric Tramway Cars', *The Times*, 6 March 1882

'Electrical Current Topics', *Electricity: A Popular Electrical Journal*, 22 July 1891

Electrician, 'The Electroscope', *New York Sun*, 29 March 1877

'Electricity a Cure for Tuberculosis', *New York Herald*, 3 August 1900

'Electricity as a Factor in Happiness', *Spectator*, 10 September 1881

'Electricity at the World's Fair', *Western Electrician*, 9 September 1893, pp. 124–5

'Electricity: The Niagara Plant', *Engineering Magazine*, 1 December 1893

Ferranti, S. Z. de, 'Pioneer of Electric Power Transmission: An Account of Some of the Early Work of Sebastian Ziani de Ferranti', *Notes and Records of the Royal Society*, 1964, 19: 33–41

Forbes, George, 'Harnessing Niagara,' *Blackwood's Magazine*, 1895, 158: 430–444

'Fruits of Genius were Swept Away', *New York Herald*, 14 March 1895

Gassiot, John Peter, 'On Some Experiments with Ruhmkorff's Induction Coil', *Philosophical Magazine*, 1854, 7: 97–99

Gassiot, John Peter, 'On the Stratification in Electrical Discharge', *Philosophical Transactions*, 1859, 149: 137

Gernsback, Hugo, 'Cold Fire', *Electrical Experimenter*, 1919, 7: 632–33

Gilliams, E. Leslie, 'Tesla's Plan of Electrically Treating School Children', *Popular Electricity Magazine*, 1912, 5: 813–14

Goldberg, Harry, 'Great Scientific Discovery Impends', *Sunday Star*, 17 May 1931

Gordon, John Edward Henry, 'The Latest Electrical Discovery,' *Nineteenth Century*, 1 March 1892, pp. 399–402

Gregory, R.A., 'Progress of Science', *Graphic*, 12 June 1897

Grove, William Robert, 'On the Progress Made in the Application of Voltaic Ignition to Lighting Mines', *Philosophical Magazine*, 1845, 27: 442–48

'Has Nikola Tesla Spoken with Mars?', *New York Journal and Advertiser*, 1 January 1901

Hawkins, Laurence A., 'Nikola Tesla, His Work and Unfulfilled Promises', *Electrical Age*, 1903, 30: 99–108

Hedley, William Snowdon, 'Apologia Pro Electricitate Suâ', *Lancet*, 1895, 145: 11050-09

Hospitalier, Édouard, 'Mr. Tesla's Experiments on Alternating Currents of Great Frequency', *Scientific American*, 26 March 1892, pp. 195–6

'Improved Rail-road Cars', *Scientific American*, 1845, 1: 1

'Influence of Inventions on Social Life', *Scientific American*, 1855, 10: 242

'Institution of Electrical Engineers', *Daily News*, 29 January 1892

Kennelly, Arthur, 'Electricity in the Household', *Scribner's Magazine*, 1890, 7: 102-15

Lardner, Dionysius, 'Railways in Ireland', *Quarterly Review*, 1839, 63: 1–34

'London, Thursday, February 4, 1892', *The Times*, 4 February 1892

Martin, Thomas Commerford, 'Tesla's Oscillator and Other Inventions', *Century Magazine*, 1895, 9: 916–33

Martin, Thomas Commerford, 'Tesla's Oscillator and other Inventions', *Century Magazine*, 1895, 27: 916–33

Martin, Thomas Commerford, 'Tesla's Oscillator and Other Inventions', *Century Magazine*, 1895, 49: 916–33

Martin, Thomas Commerford, 'Mr. Tesla and the Czar', *Electrical Engineer*, 17 November 1898

'Marvellous Ray is Produced from a New Element', *New York Journal and Advertiser*, 8 March 1901

McGovern, Chauncey Montgomery, 'The New Wizard of the West', *Pearson's Magazine*, 1899, 7: 470–76

'Mr. Tesla at the Royal Institution', *The Times*, 4 February 1892

'Mr. Tesla at Wardenclyffe', *Electrical World and Engineer*, 1901, 38: 509–10

'Mr. Tesla's Personal Exhibit at the World's Fair', *Electrical Engineer*, 1893, 16: 466–68

'Mr. Tesla's Work Destroyed by Fire', *Electrical Review*, 1895, 36: 329

Muras, T.H., 'Tesla on Energy', *Electrical Review*, 1900, 46: 1079

'Niagara Power Transmission, the Electrical Exposition, and Mr. Johnston's Libel Suit', *Western Electrician*, 30 May 1896

'Niagara's Power Transmitted to New York', *Scientific American*, 1896, 74: 283

'Nicola [*sic*] Tesla on Far Seeing', *New York Herald*, 30 August 1896

'Nikola Tesla and his Business', *New York Sun*, 24 April 1892

'Nikola Tesla at Niagara Falls', *Western Electrician*, 1 August 1896

'Nikola Tesla Discusses X Rays', *New York Times*, 11 March 1896

'Nikola Tesla Shows How Men of the Future May Become as Gods', *New York Herald*, 30 December 1900

'Nikola Tesla Talks of the Future of the Greatest Problems Now Confronting the Scientific World', *New York Press*, 2 March 1913

'Nikola Tesla's Idea of Himself', *New York Herald*, 26 November 1906

'Nikola Tesla's Predictions for 1908', *New York Post*, 5 January 1908

'Nikola Tesla's Work', *New York Sun*, 3 May 1896

Note, *The Electrician*, 19 January 1900

Porter, Rufus, 'The Scientific American', *Scientific American*, 1845, 1: 1

'Preface', *Patent Journal and Inventor's Advocate*, 1850, 10: iii–iv

'Radium', *New York Times*, 22 February 1903

Riordan, Roger, 'A Dream of the Future World's Fair', *Century Magazine*, 1901, 62: 157–58

'Scientific Discoveries', *The Star*, 4 January 1900

'Scientists Honor Nikola Tesla', *New York Herald*, 23 April 1893

Sellers, Coleman, 'The Utilization of Niagara's Power', W.D. Howells, Mark Twain and Nathaniel Shaler, *The Niagara Book* (Buffalo: Underhill and Nichols, 1893), pp. 193–220

Shaw, G.M., 'Sketch of Thomas Alva Edison', *Popular Science Monthly*, 1878, 13: 487–491

Stetson, Francis Lynde, 'The Uses of the Niagara Water Power', *Cassirer's Magazine*, 1895, 7: 173–92

Stockbridge, Frank Parker, 'Will Tesla's New Monarch of Mechanics Revolutionize the World', *Washington Post*, 15 October 1911

'Strange Light at Tesla's Tower', *New York Tribune*, 19 July 1903

'Superman of the Waldorf', *Time*, 27 November 1944

'T.C. Martin's Views', *New York Times*, 15 December 1901

Tesla, Nikola, 'A New System of Alternate Current Motors and Transformers', *Transactions of the American Institute of Electrical Engineers*, 1888, 5: 309–27

Tesla, Nikola, 'Mechanical and Electrical Oscillators', *Proceedings of the International Electrical Congress, Held in the City of Chicago, August 21st to 25th 1893* (New York: American Institute of Electrical Engineers, 1894)

Tesla, Nikola, 'Tesla's Startling Results in Radiography at Great Distances through Considerable Thickness and Substance', *Electrical Review*, 11 March 1896

Tesla, Nikola, 'High Frequency Oscillators for Electro-Therapeutic and Other Purposes', *Electrical Engineer*, 1898, 26: 477–81

Tesla, Nikola, 'Some Experiments in Tesla's Laboratory with Currents of High Potential and High Frequency', *Electrical Review*, March 1899, pp. 195–97, 204

Tesla, Nikola, 'The Problem of Increasing Human Energy,' *Century Magazine*, 1900, 38: 175-211

Tesla, Nikola, 'The Transmission of Electric Energy Without Wires', *Electrical World and Engineer*, 1904, 43: 429–31

Tesla, Nikola, 'Tesla's Tidal Wave to Make War Impossible', *New York World*, 21 April 1907

Tesla, Nikola, 'Possibilities of Wireless', *New York Times*, 22 October 1907

Tesla, Nikola, 'Tesla on Aeroplanes', *New York Times*, 15 September 1908

Tesla, Nikola, 'The True Wireless', *Electrical Experimenter*, May 1919, pp. 28–30

'Tesla and his Researches', *New York Times*, 22 January 1894

'Tesla and Telegraphy', *New York Tribune*, 27 November 1901

'Tesla and the Roentgen Rays', *New York Herald*, 23 February 1896

'Tesla Certain of his New Power', *New York Sun*, 10 July 1933

'Tesla Declares He Will Abolish War', *New York Herald*, 8 November 1898

'Tesla Has Plan to Do Away With Newspapers', *New York Herald*, 9 March 1904

'Tesla Here for Work', *Denver Rocky Mountain News*, 18 May 1899.

'Tesla Motors in Europe', *Electrical Engineer*, 28 September 1892

'Tesla on the Electric Transmission of Energy', *English Mechanic and World of Science*, 21 July 1893, pp. 44–45

'Tesla Predicts New Source of Power in Year', *New York Herald Tribune*, 9 July 1933

'Tesla Predicts Power by Radio', *New York Sun*, 23 April 1934

'Tesla Ready for Business', *New York Tribune*, 7 August 1901

'Tesla Ready for Work', *Denver Rocky Mountain News*, 20 May 1899

'Tesla Returns from Europe', *Electrical Review*, 10 September 1892

'Tesla Seeks to Send Power to the Planets', *New York Times*, 11 July 1931

'Tesla Talks and Confirms His Astounding Story', *Criterion*, 19 November 1898

'Tesla Tower to be Sold', *New York Times*, 27 October 1907

'Tesla, 75, Predicts New Power Source', *New York Times*, 5 July 1931

'Tesla's Call from Mars?' *New York Sun*, 3 January 1901

'Tesla's Experiments', *Electrical Engineer*, 11 March 1892

'Tesla's Fixtures in Sheriff's Sale', *Colorado Springs Gazette*, 10 March 1906

'Tesla's Flashes Startling', *New York Sun*, 17 July 1903

'Tesla's New War Wonder', *New York Sun*, 8 November 1898

'Tesla's System of Electrical Power Transmission through Natural Media', *Electrical Review*, 1898, 43: 709–11

'Tesla's Task of Taming Air', *Chicago Times-Herald*, 15 May 1899

'Tesla's Work at Niagara', *New York Times*, 16 July 1895

'The Advance of Electricity', *Standard*, 5 February 1892

'The Close of the Centennial Exhibition', *Scientific American*, 1876, 35: 336

'The Columbian Exposition', *Harper's Weekly*, 16 September 1893

'The Coming Force', *Punch's Almanack for 1882*, 6 December 1881

'The Crystal Palace Electrical Exhibition', *Daily News*, 17 March 1882

'The Crystal Palace Electrical Exhibition', *The Graphic*, 18 March 1882

'The Electric Eye', *Review of Reviews*, 1897, 15: 88

'The Electrical Exhibition', *Times*, 8 February 1892

'The Great Induction Coil at the Polytechnic Institution', *Times*, 7 April 1869

'The Opening of the Centennial', *Scientific American*, 1876, 34: 337.7

'The Speaker at the Crystal Palace', *Morning Post*, 29 March 1882

'The Telectroscope', *The Electrician*, 1881, 6: 141

'The Telectroscope', *The Times*, 27 January 1879

'The Tesla Electric Light Company', *Electrical Review*, 14 August 1886

'The Tesla Lecture at St Louis', *Electrical Review*, 1893, 32: 358–59

'The Tesla Turbine', *Electrical Review and Western Electrician*, 30 September 1911

'The Wizard of Menlo Park', *Daily Graphic*, 10 April 1878

'The Wizard's Search', *Daily Graphic*, 9 July 1879

'This Morning's News', *Daily News*, 10 June 1897

'To Compete with Cable Lines', *New York Times*, 15 December 1901

'To Telegraph Around the World Without Wires', *New York Herald*,
    31 January 1897

'Tributes of Former Associates', *Electrical World*, 21 March 1914, p. 637

'Trouve's Jewelry', *Electrical Review*, 27 June 1885

Twain, Mark, 'From the "London Times" of 1904', *Century Magazine*, 1898,
    57: 100–104

Viereck, George Sylvester, 'A Machine to End War', *Liberty Magazine*,
    9 February 1935, pp. 5–7

Watkins, John Elfreth, 'What May Happen in the Next Hundred Years',
    Ladies' Home Journal, 1900, 18: 8

'We May Signal to Mars', *New York Sun*, 25 March 1896

'Westinghouse Apparatus and the Tesla System Adopted at Niagara Falls',
    Electrical Engineer, 1 November 1893

'Westinghouse Work at the Fair', *Electrical Engineer*, 1893, 16: 153–54

Wetzler, Joseph, 'Electric Lamps Fed from Space, and Flames that do not
    Consume', *Harper's Weekly*, 11 July 1891

'What Electricity Will Do for Us', *Buffalo News and Sunday Express*,
    16 April 1893

'What Mr. Tesla is Said to Have Said', *Western Electrician*, 14 March 1903

'What of the Future of Electricity', *New York Herald*, 11 February 1912

Wheeler, Candace, 'A Dream City', *Harper's Magazine*, 1893, 86: 830–46

'Wireless Telegraphy Across the Atlantic', *The Times*, 16 December 1901

'Wireless Telegraphy', *Daily News*, 20 March 1899

'Wireless Telegraphy', *Morning Post*, 18 January 1900

Wynne, Arthur, 'Tesla's Latest Marvels', *New York World*, 23 February 1919

Wynter, Andrew, 'The Electric Telegraph', *Quarterly Review*, 1854, 95:
    118–64

Notes

Chapter 1: A Child of the Storm

1. For details of Tesla's background and childhood see the excellent W. Bernard Carlson, *Tesla: Inventor of the Electrical Age* (Princeton: Princeton University Press, 2013). For a history of the Austrian Empire see Pieter Judson, *The Habsburg Empire: A New History* (Cambridge MA: Harvard University Press, 2016).

2. The story is repeated in many places. See for example Dan Mrkich, *Tesla: The European Years* (Ottawa: Commoners Publishing, 2010), p. 55.

3. Nikola Tesla, *My Inventions and Other Writings* (London: Penguin Books, 2011), p. 8. The original 'My Inventions' appeared as a series of articles in the magazine *Electrical Experimenter* in 1919, commissioned by the editor Hugo Gernsback.

4. Mrkich, op. cit., note 2, p. 54.

5. [Dionysius Lardner], 'Railways in Ireland', *Quarterly Review*, 1839, 63: 1–34, on p. 13.

6. Quoted in Carlson, op. cit., note 1, p. 18.

7. Tesla, *My Inventions*, op. cit., note 3, p. 25.

8. Tesla, *My Inventions*, op. cit., note 3, p. 6.

9. Tesla, *My Inventions*, op. cit., note 3, p. 28.

10. Tesla, *My Inventions*, op. cit., note 3, p. 31.

11. Tesla, *My Inventions*, op. cit., note 3, p. 32.

12. Tesla, *My Inventions*, op. cit., note 3, p. 32.

13. Tesla, *My Inventions*, op. cit., note 3, p. 34.

14. Tesla, *My Inventions*, op. cit., note 3, p. 35.

15. Tesla, *My Inventions*, op. cit., note 3, p. 34.

Chapter 2: Electric Power

1. Charles Grindriez and James M. Hart, *International Exhibitions: Paris–Philadelphia–Vienna* (New York: A.S. Barnes & Co., 1878), p. 22.

2. Ibid., p. 26.

3. Clement Shorter, *The Brontës: Life and Letters* (London: Hodder & Stoughton, 1908), vol. 2, p. 216.

4. Quoted in K.G. Beauchamp, *Exhibiting Electricity* (London: Institution of Electrical Engineers, 1997), p. 98.

5. Alfred Smee, *Elements of Electrometallurgy*, 3rd edn (London, 1844), p. 348.

6. Henry M. Noad, *A Course of Eight Lectures on Electricity, Galvanism, Magnetism, and Electro-magnetism* (London, 1839), pp. 381–2.

7. 'Davenport's Electro-magnetic Engine', *Mechanics' Magazine*, 1837, 27: 404–5.

8. See Iwan Rhys Morus, *William Robert Grove: Victorian Gentleman of Science* (Cardiff: University of Wales Press, 2017).

9. Quoted in Robert Post, *Physics, Patents and Politics: A Biography of Charles Grafton Page* (New York: Science History Publications, 1976), p. 98.

10. William Robert Grove, *The Correlation of Physical Forces*, 6th edn (London: Longmans, Green and Co., 1874), p. 140.

11. Nicholas Callan, 'On a New Galvanic Battery', *Philosophical Magazine*, 1836, 9: 472–78, on p. 478.

12. John Peter Gassiot, 'On Some Experiments with Ruhmkorff's Induction Coil', *Philosophical Magazine*, 1854, 7: 97–99, on p. 99.

13. John Peter Gassiot, 'On the Stratification in Electrical Discharge', *Philosophical Transactions*, 1859, 149: 137.

14. 'The Great Induction Coil at the Polytechnic Institution', *Times*, 7 April 1869, p. 4. For the Polytechnic Institution see Brenda Weeden, *The Education of the Eye: History of the Polytechnic Institution, 1838–1881* (Cambridge, UK, 2008).

15. William Robert Grove, 'On the Progress Made in the Application of Voltaic Ignition to Lighting Mines', *Philosophical Magazine*, 1845, 27: 442–48.

16. Quoted in Wolfgang Schivelbusch, *Disenchanted Night: The Industrialization of Light* (Berkeley, London and Los Angeles: University of California Press, 1988), p. 55.

17. 'Electric Light', *Patent Journal*, 1849, 6: 80.

18. Quoted in Schivelbusch, op. cit., note 16, p. 55.

19. Iwan Rhys Morus, *Shocking Bodies* (Stroud: History Press, 2011), p. 116.

20. Ibid., p. 114.

21. Beauchamp, op. cit., note 4, p. 127.

22. Beauchamp, op. cit., note 4, p. 129.

Chapter 3: Working Electricity

1. Latimer Clark, 'Inaugural Address', *Journal of the Society of Telegraph Engineers*, 1875, 4: 1–22, on p. 2.

2. 'Preface', *Patent Journal and Inventor's Advocate*, 1850, 10: iii–iv, on p. iv.

3. Geoffrey Hubbard, *Cooke, Wheatstone and the Invention of the Electric Telegraph* (London: Routledge & Kegan Paul, 1965).

4. Edward Lind Morse, *Samuel F.B. Morse: His Letters and Journals*, 2 volumes (Boston: Houghton Mifflin, 1914) vol. 2, p. 6.

5. Paul Israel, *Edison: A Life of Invention* (New York: John Wiley, 1998).

6. Jeffrey Kieve, *The Electric Telegraph: A Social and Economic History* (Newton Abbott: David & Charles, 1973).

7. Daniel Headrick, *The Invisible Weapon: Telecommunications and National Politics, 1851–1945* (Oxford: Oxford University Press, 1991).

8. Andrew Wynter, 'The Electric Telegraph', *Quarterly Review*, 1854, 95: 118–64.

9. Crosbie Smith and M. Norton Wise, *Energy and Empire: A Biographical Study of Lord Kelvin* (Cambridge: Cambridge University Press, 1989), pp. 661–6.

10. David Cahan, *An Institute for an Empire: The Physikalisch-Technische Reichsanstalt 1871–1918* (Cambridge: Cambridge University Press, 1989); Iwan Rhys Morus, *When Physics Became King* (Chicago: University of Chicago Press, 2005), pp. 226–60.

11. Tesla, *My Inventions*, p. 37.

12. Carlson, p. 46.

13. Tesla, *My Inventions*, p. 43.

Chapter 4: A New World

1. Tesla, *My Inventions*, p. 48.

2. Charles Coleman Sellers, *Mr. Peale's Museum: Charles Willson Peale and the First Popular Museum of Natural Science and Art* (New York: Norton, 1980).

3. Iwan Rhys Morus, *Frankenstein's Children: Electricity, Exhibition and Experiment in Early Nineteenth-Century London* (Princeton: Princeton University Press, 1998).

4. James Cook, *The Arts of Deception* (Cambridge MA: Harvard University Press, 2001).

5. 'Astounding News!' *New York Sun*, 13 April 1844.

6. Albert Moyer, *Joseph Henry: The Rise of an American Scientist* (Washington DC: Smithsonian Institution Press, 1997).

7. Walter Rice Davenport, *Thomas Davenport: Pioneer Inventor* (Montpelier: Vermont Historical Society, 1929).

8. Rufus Porter, 'The Scientific American', *Scientific American*, 1845, 1: 1.

9. 'Improved Rail-road Cars', *Scientific American*, 1845, 1: 1.

10. 'Influence of Inventions on Social Life', *Scientific American*, 1855, 10: 242.

11. 'The Opening of the Centennial', *Scientific American*, 1876, 34: 337.

12. 'The Close of the Centennial Exhibition', *Scientific American*, 1876, 35: 336.

13. Edward Bulwer Lytton, *The Coming Race* (London: Backwoods, 1871).

14. William Livingston Alden, *Van Wagener's Ways* (London: C. Arthur Pearson Ltd, 1898).

15. Electrician, 'The Electroscope', *New York Sun*, 29 March 1877.

16. George R. Carey, 'Seeing by Electricity', *Scientific American*, 1880, 42: 355.

17. George Beard, *American Nervousness* (New York: G.P. Putnam's Sons, 1881), p. vii.

18. Ibid., p. 13.

Chapter 5: The Wizard of Menlo Park

1. Tesla, *My Inventions*, p. 48.

2. Ibid., p. 49.

3. 'The Wizard's Search', *Daily Graphic*, 9 July 1879, p. 1; 'The Wizard of Menlo Park', *Daily Graphic*, 10 April 1978.
4. Israel, *Edison*, pp. 5–8.
5. Frank Dyer and Thomas Commerford Martin, *Edison: His Life and Inventions* (New York: Harper & Brothers, 1910), p. 48.
6. Israel, op. cit., note 4, pp. 11–12.
7. Dyer and Martin, op. cit., note 5, p. 54.
8. Ibid., p. 140.
9. Ibid., p. 205.
10. Ibid., p. 209.
11. G.M. Shaw, 'Sketch of Thomas Alva Edison', *Popular Science Monthly*, 1878, 13: 487–491, on p. 489.

Chapter 6: AC/DC

1. Nikola Tesla, 'A New System of Alternate Current Motors and Transformers', *Transactions of the American Institute of Electrical Engineers*, 1888, 5: 309–27, on p. 327.
2. Ibid., p. 324.
3. Ibid., p. 326.
4. Ibid., p. 327.
5. 'The Tesla Electric Light Company', *Electrical Review*, 14 August 1886, p. 1.
6. Quoted in Carlson, p. 80.
7. Post, *Physics*.
8. Israel, *Edison*, pp. 167–90.
9. Ibid., p. 206.
10. Quoted in ibid., p. 214.
11. Quoted in S.Z. de Ferranti, 'Pioneer of Electric Power Transmission: An Account of Some of the Early Work of Sebastian Ziani de Ferranti', *Notes and Records of the Royal Society*, 1964, 19: 33–41, on p. 37.
12. Quoted in ibid., p. 37.

Chapter 7: Building Tomorrow

1. Smith and Wise, *Energy and Empire*, pp. 670–72.
2. Thomas Hughes, *Networks of Power: Electrification in Western Society, 1880–1930* (Baltimore MD: Johns Hopkins University Press, 1983), p. 241. The original cartoon was in *Electrical Plant*, May 1889.
3. 'The Coming Force', *Punch's Almanack for 1882*, 6 December 1881.
4. 'The Crystal Palace Electrical Exhibition', *Daily News*, 17 March 1882, p. 2.
5. 'The Speaker at the Crystal Palace', *Morning Post*, 29 March 1882, p. 3.
6. 'The Crystal Palace Electrical Exhibition', *The Graphic*, 18 March 1882.
7. 'Electric Tramway Cars', *The Times*, 6 March 1882, p. 6.

8. Gijs Mom, *The Electric Vehicle: Technology and Expectations in the Automobile Age* (Baltimore MD: Johns Hopkins University Press, 2004), p. 21.

9. 'The Telectroscope', *The Times*, 27 January 1879

10. 'The Telectroscope', *The Electrician*, 1881, 6: 141.

11. 'Electricity as a Factor in Happiness', *Spectator*, 10 September 1881, p. 9.

12. Jules Verne, *Twenty Thousand Leagues Under the Sea*, first published 1869, this edn trans. Mendor T. Brunetti (Harmondsworth: Penguin, 1994), p. 86.

13. Albert Robida, *The Twentieth Century*, trans. Philippe Willems (Middletown CT: Wesleyan University Press, 2004). The telephonoscope first appears on p. 50.

14. Steve Carper, 'The Mighty Electrical Men', https://www.blackgate.com/2018/02/07/the-mighty-electric-men/ (accessed 29/02/2019).

15. Iwan Rhys Morus, *When Physics Became King* (Chicago: University of Chicago Press, 2005), pp. 253–59.

16. Daniel Headrick, *The Invisible Weapon: Telecommunications and National Politics, 1851–1945* (Oxford: Oxford University Press, 1991).

17. W. Bernard Carlson, 'Elihu Thomson: Man of Many Facets', *IEEE Spectrum*, 1983, 20: 72–75.

Chapter 8: The Business of Invention

1. 'Tributes of Former Associates', *Electrical World*, 21 March 1914, p. 637.

2. Carlson, p. 112.

3. *Reports of the United States Commissioners to the Universal Exposition of 1889 at Paris* (Washington DC: Government Printing Office, 1891), vol. 4, p. 13.

4. Ibid., p. 44.

5. Mark Essig, *Edison and the Electric Chair* (Stroud: Sutton Publishing, 2003).

6. Harold P. Brown, *The Comparative Danger to Life of the Alternating and Continuous Electrical Currents* (New York, 1889).

7. *New York Times*, 7 August 1890, p. 1.

8. James Clerk Maxwell, *A Treatise on Electricity and Magnetism*, 2 vols (Cambridge: Cambridge University Press, 1873), vol. 2, p. 493.

9. Nikola Tesla, 'The True Wireless', *Electrical Experimenter*, May 1919, pp. 28–30, on p. 28.

10. Nikola Tesla, 'Some Experiments in Tesla's Laboratory with Currents of High Potential and High Frequency', *Electrical Review*, March 1899, pp. 195–97, 204, on p. 195.

Chapter 9: Electrical Landscapes

1. 'A Town Lighted by Electricity', *Scientific American*, May 1880, p. 275.

2. Graeme Gooday, *Domesticating Electricity: Technology, Uncertainty and Gender, 1880–1914* (London: Routledge, 2008).

3. Arthur Kennelly, 'Electricity in the Household', *Scribner's Magazine*, 1890, 7: 102–15, on p. 115.
4. 'Electric Light News', *Electrical Review*, 16 October 1886, p. 9.
5. Carolyn Marvin, *When Old Technologies Were New: Thinking About Electrical Communication in the Late Nineteenth Century* (Oxford: Oxford University Press, 1988), p. 160.
6. 'Trouve's Jewelry', *Electrical Review*, 27 June 1885, p. 2.
7. Both quotes from Marvin, op. cit., note 5, pp. 138–9.
8. Oliver Lodge, *Past Years* (London: Hodder & Stoughton, 1931), p. 185.
9. Iwan Rhys Morus, *Michael Faraday and the Electrical Century* (London: Icon Books, 2005).
10. Morus, *Shocking Bodies*, pp. 20–38.
11. George Beard, *American Nervousness* (New York: G.P. Putnam's Sons, 1881).
12. 'Experiments with Alternate Currents of Very High Frequency and their Application to Methods of Artificial Illumination', Thomas Commerford Martin (ed.), *The Inventions, Researches and Writings of Nikola Tesla*, 2nd edn (New York: Barnes & Noble, 1995), pp. 145–97, on p. 146.
13. Ibid., p. 188.
14. 'Electrical Current Topics', *Electricity: A Popular Electrical Journal*, 22 July 1891, p. 7.
15. Joseph Wetzler, 'Electric Lamps Fed from Space, and Flames that do not Consume', *Harper's Weekly*, 11 July 1891, p. 524.
16. E.C. Baker, *Sir William Preece F.R.S.: Victorian Engineer Extraordinary* (London: Hutchinson, 1976).
17. 'Institution of Electrical Engineers', *Daily News*, 29 January 1892, p. 3.
18. 'Experiments with Alternate Currents of High Potential and High Frequency', Thomas Commerford Martin (ed.), *The Inventions, Researches and Writings of Nikola Tesla*, 2nd edn (New York: Barnes & Noble, 1995), pp. 198–293, on p. 199.
19. Ibid., pp. 200–1.
20. 'The Advance of Electricity', *Standard*, 5 February 1892, p. 3.
21. 'Mr. Tesla at the Royal Institution', *The Times*, 4 February 1892, p. 6.
22. 'London, Thursday, February 4, 1892', *The Times*, 4 February 1892, p. 9.
23. 'The Electrical Exhibition', *Times*, 8 February 1892, p. 9.
24. 'Crystal Palace – Electrical Exhibition', *Morning Post*, 19 April 1892, p. 1.
25. 'Tesla's Experiments', *Electrical Engineer*, 11 March 1892, p. 242.
26. Édouard Hospitalier, 'Mr. Tesla's Experiments on Alternating Currents of Great Frequency', *Scientific American*, 26 March 1892, pp. 195–6, on p. 196.
27. Nikola Tesla, 'The True Wireless', *Electrical Experimenter*, May 1919, pp. 28–30, on p. 28.
28. 'Tesla Returns from Europe', *Electrical Review*, 10 September 1892, p. 31.

29. 'Tesla Motors in Europe', *Electrical Engineer*, 28 September 1892, p. 291.
30. Nikola Tesla, *My Inventions*, p. 58.

Chapter 10: Harnessing Nature

1. Tesla, *My Inventions*, p. 28.
2. John Edward Henry Gordon, 'The Latest Electrical Discovery,' *Nineteenth Century*, 1 March 1892, pp. 399–402, on p. 402.
3. William Pole, *The Life of Sir William Siemens* (London: John Murray, 1888), p. 249.
4. Ibid., p. 250.
5. Smith and Wise, *Energy and Empire*, p. 714.
6. Quoted in Carlson, p. 167.
7. Quoted in Carlson, pp. 168–9.
8. Quoted in Carlson, p. 171.
9. Coleman Sellers, 'The Utilization of Niagara's Power', W.D. Howells, Mark Twain and Nathaniel Shaler, *The Niagara Book* (Buffalo: Underhill and Nichols, 1893), pp. 193–220, on p. 193.
10. George Forbes, 'Harnessing Niagara,' *Blackwood's Magazine*, 1895, 158: 430–444, on pp. 443–4.
11. Sellers, op. cit., note 9, p. 219.
12. Sellers, op. cit., pp. 217–18.
13. Patrick McGreevy, *Imagining Niagara: The Meaning and Making of Niagara Falls* (Amherst: University of Massachusetts Press, 2009), quote on p. 82.
14. Francis Lynde Stetson, 'The Uses of the Niagara Water Power', *Cassirer's Magazine*, 1895, 7: 173–92, on pp. 173–4.
15. Silvanus P. Thompson, *Life of Lord Kelvin*, vol. 2 (London: Macmillan, 1910), p. 1002.
16. E. Jay Edwards, 'The Capture of Niagara', *McClure's Magazine*, 1894, 3: 423–35, on p. 435.
17. 'What Electricity Will Do for Us', *Buffalo News and Sunday Express*, 16 April 1893, pp. 2–3, on p. 2.
18. 'Capture', op. cit. note 16, p. 426.
19. 'Westinghouse Apparatus and the Tesla System Adopted at Niagara Falls', *Electrical Engineer*, 1 November 1893, p. 399; 'Electricity: The Niagara Plant', *Engineering Magazine*, 1 December 1893, p. 358.
20. 'Tesla's Work at Niagara', *New York Times*, 16 July 1895, p. 10.
21. 'Nikola Tesla at Niagara Falls', *Western Electrician*, 1 August 1896, p. 55.
22. 'Niagara's Power Transmitted to New York', *Scientific American*, 1896, 74: 283.
23. 'Niagara Power Transmission, the Electrical Exposition, and Mr. Johnston's Libel Suit', *Western Electrician*, 30 May 1896, p. 271.

Chapter 11: The Greatest Show on Earth

1. Trumbull White and William Ingleheart, *The World's Columbian Exposition* (Boston: John K. Hastings, 1893), p. 305.

2. Ibid., p. 304.

3. Ibid., p. 43.

4. Ibid., pp. 50–1.

5. Candace Wheeler, 'A Dream City', *Harper's Magazine*, 1893, 86: 830–46, on p. 830.

6. Ibid., p. 846.

7. 'At the Fair', *Century Magazine*, 1893, 46: 3–21, on p. 6; p. 7.

8. 'The Columbian Exposition', *Harper's Weekly*, 16 September 1893, p. 878.

9. White and Ingleheart, op. cit., note 1, p. 314.

10. Ibid., p. 315

11. Ibid., p. 316.

12. Ibid., p. 319.

13. Ibid., p. 327.

14. Ibid., p. 318.

15. Ibid., p. 320.

16. Ibid., p. 322.

17. 'Mr. Tesla's Personal Exhibit at the World's Fair', *Electrical Engineer*, 1893, 16: 466–68, on p. 466.

18. Ibid., p. 468.

19. White and Ingleheart, op. cit., note 1, p. 319.

20. 'Westinghouse Work at the Fair', *Electrical Engineer*, 1893, 16: 153–54, on p. 153.

21. White and Ingleheart, op. cit., note 1, p. 301.

22. Ibid., p. 301.

23. Ibid., p. 306.

24. Ibid., pp. 309–10.

25. Nikola Tesla, 'Mechanical and Electrical Oscillators', *Proceedings of the International Electrical Congress, Held in the City of Chicago, August 21st to 25th 1893* (New York: American Institute of Electrical Engineers, 1894) pp. 475–82.

26. 'Electricity at the World's Fair', *Western Electrician*, 9 September 1893, pp. 124–5, on p. 124.

27. Thomas Commerford Martin, 'Tesla's Oscillator and Other Inventions', *Century Magazine*, 1895, 49: 916–33, on p. 916.

Chapter 12: In the Ether

1. William Crookes, 'Electricity in Relation to Science', *Popular Science Monthly*, 1892, 40: 497–500, on p. 498.

2. William Crookes, 'Some Possibilities of Electricity', *Fortnightly Review*, 1892, 51: 173–81, on p. 176.

3. Ibid., p. 176.
4. Ibid., p. 177.
5. Oliver Lodge, *Modern Views of Electricity* (London: Macmillan, 1889), p. viii.
6. Ibid., p. xii.
7. Oliver Lodge, *Modern Views of Electricity*, 2nd edition (London: Macmillan, 1892), p. 356.
8. Oliver Lodge, *Past Years: An Autobiography* (London: Hodder & Stoughton, 1931), p. 230.
9. Ivor Hughes and David Ellis Evans, *Before We Went Wireless: David Edward Hughes FRS, His Life, Inventions and Discoveries* (Benington VT: Images from the Past, 2011).
10. J.J. Fahie, *A History of Wireless Telegraphy* (London: Blackwoods, 1899), pp. 147–8.
11. Ibid., p. 151.
12. Quoted from the recollections of P.R. Mullis in E.C. Baker, *Sir William Preece F.R.S.: Victorian Engineer Extraordinary* (London: Hutchinson, 1976), pp. 266–7.
13. Fahie, op. cit., note 10, p. 316.
14. Quoted in Baker, op. cit., note 11, pp. 269–70.
15. Fahie, op. cit., note 10, p. 218.
16. 'This Morning's News', *Daily News*, 10 June 1897, p. 1.
17. R.A. Gregory, 'Progress of Science', *Graphic*, 12 June 1897, p. 730.
18. Quoted in Fahie, op. cit., note 10, p. 219.
19. Ibid., p. 227.
20. 'Wireless Telegraphy', *Daily News*, 20 March 1899, p. 5.
21. Ibid.
22. 'To Telegraph Around the World Without Wires', *New York Herald*, 31 January 1897, p. 1.
23. 'Tesla on the Electric Transmission of Energy', *English Mechanic and World of Science*, 21 July 1893, pp. 44–5, on p. 44.

Chapter 13: The Pursuit of Power

1. Tesla, *My Inventions*, p. 58.
2. 'On Light and Other High Frequency Currents', Thomas Commerford Martin (ed.), *The Inventions, Researches and Writings of Nikola Tesla*, 2nd edn (New York: Barnes & Noble, 1995), pp. 294–373, on p. 301.
3. Ibid., p. 346.
4. 'The Tesla Lecture at St Louis', *Electrical Review*, 1893, 32: 358–9, on p. 359.
5. Thomas Commerford Martin, 'Tesla's Oscillator and Other Inventions', *Century Magazine*, 1895, 9: 916–33, on p. 916.
6. Ibid., p. 933.

7. Ibid., p. 931.
8. 'Tesla's System of Electrical Power Transmission through Natural Media', *Electrical Review*, 1898, 43: 709–11, on pp. 709–10.
9. 'Tesla Here for Work', *Denver Rocky Mountain News*, 18 May 1899.
10. 'Tesla Ready for Work', *Denver Rocky Mountain News*, 20 May 1899.
11. Chauncey Montgomery McGovern, 'The New Wizard of the West', *Pearson's Magazine*, 1899, 7: 470–6, on p. 470.
12. Carlson, pp. 267–8.
13. Nikola Tesla, 'The Transmission of Electric Energy Without Wires', *Electrical World and Engineer*, 1904, 43: 429–31, on p. 429.
14. Ibid., p. 430.
15. Nikola Tesla, 'The Problem of Increasing Human Energy', *Century Magazine*, 1900, 38: 175–211. Republished in Nikola Tesla, *My Inventions*, pp. 101–67.
16. Peter Bowler, *The Invention of Progress* (Oxford; Blackwells, 1989).
17. Tesla, *My Inventions*, pp. 104–5.
18. Tesla, *My Inventions*, pp. 107–10.
19. Tesla, *My Inventions*, p. 118.
20. T.H. Muras, 'Tesla on Energy', *Electrical Review*, 1900, 46: 1079.
21. Editorial, *Electrician*, 1900, 45: 274.

Chapter 14: Other Worlds

1. 'Nikola Tesla and his Business', *New York Sun*, 24 April 1892.
2. 'Scientists Honor Nikola Tesla', *New York Herald*, 23 April 1893.
3. 'Tesla and his Researches', *New York Times*, 22 January 1894.
4. Thomas Commerford Martin, 'Tesla's Oscillator and other Inventions', *Century Magazine*, 1895, 27: 916–33.
5. 'Fruits of Genius were Swept Away', *New York Herald*, 14 March 1895.
6. Editorial, *New York Sun*, 14 March 1895.
7. 'Mr. Tesla's Work Destroyed by Fire', *Electrical Review*, 1895, 36: 329 (15 March)
8. 'Nikola Tesla's Work', *New York Sun*, 3 May 1896.
9. Francis Leon Chrisman, 'Our Modern Franklin – Nikola Tesla', *Success*, 4 November 1898, p. 136.
10. 'Wireless Telegraphy', *Daily News*, 20 March 1899, p. 5.
11. *Town Topics*, 6 April 1899, p. 10.
12. 'Nikola Tesla Discusses X Rays', *New York Times*, 11 March 1896.
13. 'Tesla and the Roentgen Rays', *New York Herald*, 23 February 1896.
14. 'A Submarine Destroyer that Really Destroys', *New York Journal*, 13 November 1898.
15. 'Tesla Talks and Confirms His Astounding Story', *Criterion*, 19 November 1898.

16. 'Tesla's New War Wonder', *New York Sun*, 8 November 1898.
17. 'Tesla Declares He Will Abolish War', *New York Herald*, 8 November 1898.
18. 'Doubts Value of Tesla Discovery', *New York Herald*, 9 November 1898.
19. Thomas Commerford Martin, 'Mr. Tesla and the Czar', *Electrical Engineer*, 17 November 1898, pp. 486–7, on p. 487.
20. Nikola Tesla, 'The Problem of Increasing Human Energy', *My Inventions*, p. 115.
21. 'Nicola [*sic*] Tesla on Far Seeing', *New York Herald*, 30 August 1896.
22. 'We May Signal to Mars', *New York Sun*, 25 March 1896.
23. 'Tesla's Task of Taming Air', *Chicago Times-Herald*, 15 May 1899.
24. 'Has Nikola Tesla Spoken with Mars?', *New York Journal and Advertiser*, 1 January 1901.
25. 'Tesla's Call from Mars?' *New York Sun*, 3 January 1901.

Chapter 15: Wardenclyffe

1. Note, *The Electrician*, 19 January 1900, p. 423.
2. Carlson, p. 458n.
3. 'A Tesla Patent', *Electrical World and Engineer*, 31 March 1900, p. 477.
4. 'Electricity a Cure for Tuberculosis', *New York Herald*, 3 August 1900.
5. '500,000 Volts of Electricity Passed Through Body to Cure Consumption – Tesla's Idea', *New York World*, 19 August 1900.
6. 'Nikola Tesla Shows How Men of the Future May Become as Gods', *New York Herald*, 30 December 1900.
7. 'Wireless Telegraphy', *Morning Post*, 18 January 1900, p. 3.
8. 'Scientific Discoveries', *The Star*, 4 January 1900, p. 1.
9. 'A Hungarian Wizard', *Pall Mall Gazette*, 22 June 1900, p. 4.
10. Quoted in Carlson, p. 304.
11. Ron Chernow, *The House of Morgan: An American Banking Dynasty and the Rise of Modern Finance* (New York: Grove Press, 2010).
12. Tesla to Morgan, 10 December 1900, quoted in Carlson, p. 314.
13. Editorial, *Western Electrician*, 9 March 1901.
14. 'Tesla Ready for Business', *New York Tribune*, 7 August 1901.
15. 'Mr. Tesla at Wardenclyffe', *Electrical World and Engineer*, 1901, 38: 509–10, on p. 510.
16. 'Tesla and Telegraphy', *New York Tribune*, 27 November 1901.
17. 'Inventor Tesla's Plant Nearing Completion', *Brooklyn Eagle*, 7 February 1902.
18. Carlson, p. 326.
19. 'Cloudborn Electric Wavelets to Encircle the Globe', *New York Times*, 27 March 1904.
20. 'What Mr. Tesla is Said to Have Said', *Western Electrician*, 14 March 1903, p. 211.

21. 'Tesla's Flashes Startling', *New York Sun*, 17 July 1903.

22. 'Strange Light at Tesla's Tower', *New York Tribune*, 19 July 1903.

23. Laurence A. Hawkins, 'Nikola Tesla, His Work and Unfulfilled Promises', *Electrical Age*, 1903, 30: 99–108, on p. 99, p. 102, p. 105.

24. 'Wireless Telegraphy Across the Atlantic', *The Times*, 16 December 1901, p. 5.

25. 'T.C. Martin's Views', *New York Times*, 15 December 1901.

26. 'To Compete with Cable Lines', *New York Times*, 15 December 1901.

27. Tesla to Morgan, 9 January 1902, quoted in Carlson, p. 339.

28. Nikola Tesla, 'The Transmission of Electrical Energy Without Wires', *Electrical World and Engineer*, 1904, 43: 429–31, on p. 430.

29. 'Tesla Has Plan to Do Away With Newspapers', *New York Herald*, 9 March 1904.

30. Tesla, op. cit., note 27, p. 431.

31. 'Nikola Tesla's Idea of Himself', *New York Herald*, 26 November 1906.

Chapter 16: Inventing the Future

1. John Jacob Astor, *A Journey in Other Worlds: A Romance of the Future* (New York: D. Appleton and Co., 1894), p. 35.

2. Ibid., p. 48.

3. George Griffith, 'The Raid of *Le Vengeur*', reprinted from *Pearson's Magazine* (1901) in I.F. Clarke (ed.), *The Tale of the Next Great War, 1871–1914* (Liverpool: Liverpool University Press, 1995), pp. 193–209, on p. 193.

4. George Griffith, *The Angel of the Revolution: A Tale of the Coming Terror* (London: Tower Publishing Company, 1893), p. 1.

5. L.J. Beeston, 'A Star Fell', reprinted from *Cassell's Magazine* in *Science Fiction by the Rivals of H.G. Wells* (Secaucus, NJ: Castle Books, 1979), pp. 25–33.

6. E.E. Kellett, 'The Lady Automaton', reprinted from *Pearson's Magazine* in *Science Fiction by the Rivals of H.G. Wells* (Secaucus, NJ: Castle Books, 1979), pp. 351–63.

7. Will Tattersdill, *Science, Fiction, and the Fin-de-Siècle Periodical Press* (Cambridge: Cambridge University Press, 2016).

8. Mark Twain, 'From the "London Times" of 1904', *Century Magazine*, 1898, 57: 100–104.

9. Roger Riordan, 'A Dream of the Future World's Fair', *Century Magazine*, 1901, 62: 157–58, on p. 158.

10. H.G. Wells, *Anticipations of the Reaction of Mechanical and Scientific Progress upon Human Life and Thought* (London: Chapman & Hall, 1902), p. 241.

11. Charles Gibson, *The Romance of Modern Electricity* revised edition (Philadelphia: J.B. Lippincott, 1910), pp. 6–7.

12. Richard Kerr, *Wireless Telegraphy, Popularly Explained* (New York: Charles Scribner's Sons, 1898), p. 7.

13. John Elfreth Watkins, 'What May Happen in the Next Hundred Years', *Ladies' Home Journal*, 1900, 18: 8.

14. Quoted in Catherine Caufield, *Multiple Exposures: Chronicles of the Radiation Age* (Harmondsworth: Penguin, 1989), p. 7.

15. Nikola Tesla, 'Tesla's Startling Results in Radiography at Great Distances through Considerable Thickness and Substance', *Electrical Review*, 11 March 1896, pp. 131–5, on p. 135.

16. William Snowdon Hedley, 'Apologia Pro Electricitate Suâ', *Lancet*, 1895, 145: 11050-09.

17. 'The Electric Eye', *Review of Reviews*, 1897, 15: 88.

18. US Patent 755840A, Jagadish Chandra Bose, *Detector for Electrical Disturbances*, filed 30 September 1901, granted 29 March 1904.

19. 'Radium', *New York Times*, 22 February 1903.

20. 'Marvellous Ray is Produced from a New Element', *New York Journal and Advertiser*, 8 March 1901.

Chapter 17: Projections

1. 'Tesla's Fixtures in Sheriff's Sale', *Colorado Springs Gazette*, 10 March 1906.

2. Nikola Tesla, 'Possibilities of Wireless', *New York Times*, 22 October 1907.

3. 'Tesla Tower to be Sold', *New York Times*, 27 October 1907.

4. Carlson, p. 379.

5. 'Nikola Tesla's Predictions for 1908', *New York Post*, 5 January 1908.

6. 'What of the Future of Electricity', *New York Herald*, 11 February 1912.

7. 'Nikola Tesla Talks of the Future of the Greatest Problems Now Confronting the Scientific World', *New York Press*, 2 March 1913.

8. Arthur Wynne, 'Tesla's Latest Marvels', *New York World*, 23 February 1919, p. 6.

9. Nikola Tesla, 'Tesla's Tidal Wave to Make War Impossible', *New York World*, 21 April 1907, pp. 6–7, on p. 7.

10. Nikola Tesla, 'Tesla on Aeroplanes', *New York Times*, 15 September 1908.

11. Frank Parker Stockbridge, 'Will Tesla's New Monarch of Mechanics Revolutionize the World', *Washington Post*, 15 October 1911.

12. 'The Tesla Turbine', *Electrical Review and Western Electrician*, 30 September 1911, p. 673.

13. Stockbridge, op. cit., note 11.

14. Nikola Tesla, 'High Frequency Oscillators for Electro-Therapeutic and Other Purposes', *Electrical Engineer*, 1898, 26: 477–81, on p. 477.

15. E. Leslie Gilliams, 'Tesla's Plan of Electrically Treating School Children', *Popular Electricity Magazine*, 1912, 5: 813–14, on p. 814.

16. Hugo Gernsback, 'Cold Fire', *Electrical Experimenter*, 1919, 7: 632–33, on p. 633.

17. Harry Goldberg, 'Great Scientific Discovery Impends', *Sunday Star*, 17 May 1931, p. 3.

18. 'Tesla, 75, Predicts New Power Source', *New York Times*, 5 July 1931, p. 3.
19. 'Tesla Seeks to Send Power to the Planets', *New York Times*, 11 July 1931.
20. 'Tesla Predicts New Source of Power in Year', *New York Herald Tribune*, 9 July 1933.
21. 'Tesla Certain of his New Power', *New York Sun*, 10 July 1933.
22. 'Tesla Predicts Power by Radio', *New York Sun*, 23 April 1934.
23. 'Beam to Kill Army at 200 Miles', *New York Herald Tribune*, 11 July 1934.
24. '2,000 Are Present at Tesla Funeral', *New York Times*, 13 January 1943, p. 24.

Chapter 18: The Afterlives of Nikola Tesla

1. George Sylvester Viereck, 'A Machine to End War', *Liberty Magazine*, 9 February 1935, pp. 5–7, on p. 6.
2. 'Tesla, 75, Predicts New Power Source', *New York Times*, 5 July 1931.
3. John O'Neill, *Prodigal Genius: The Life of Nikola Tesla* (New York: Ives Washburn, 1944).
4. 'Superman of the Waldorf', *Time*, 27 November 1944, pp. 90–2.

Acknowledgements

My fascination with Nikola Tesla has its origins in my long-standing interest in the relationship between science and spectacle, and more recently in my increasing conviction not only that the future has a history, but that understanding that history matters a great deal for all our futures now. I have had far too many useful and fascinating conversations along the way with far too many friends and colleagues to be able to thank them all. I would, however, like to thank some of them in particular. Countless conversations with my wife and colleague Amanda Rees about the history of the future and why it matters have been absolutely indispensable. Simon Schaffer has been an inspiration throughout my career as a historian of science. Discussions with Rob Iliffe and the late Jeff Hughes over the years have been vital in helping me think more clearly about the history of science. Some of my recent thoughts about why the history of the future matters have had an airing online through *Aeon*, and I would like to thank Marina Benjamin for that opportunity.

Thanks are also due to everyone at Icon Books who have helped to make this book happen. I am especially grateful to Andrew Furlow, Duncan Heath, Victoria Reed and (particularly) Robert Sharman.

Index